Environmental Monitoring

Environmental Monitoring

Clara Simon

MURPHY & MOORE
www.murphy-moorepublishing.com

Murphy & Moore Publishing,
1 Rockefeller Plaza,
New York City, NY 10020, USA

Visit us on the World Wide Web at:
www.murphy-moorepublishing.com

ISBN: 978-1-63987-195-7 (Hardback)

Cataloging-in-Publication Data

Environmental monitoring / Clara Simon.
p. cm.
Includes bibliographical references and index.
ISBN 978-1-63987-195-7
1. Environmental monitoring. 2. Environmental engineering. 3. Pollution--Measurement.
I. Simon, Clara.
TD193 .E58 2022
363.706 3--dc23

Table of Contents

Preface

This book has been written, keeping in view that students want more practical information. Thus, my aim has been to make it as comprehensive as possible for the readers. I would like to extend my thanks to my family and co-workers for their knowledge, support and encouragement all along.

The process of sampling, characterizing and monitoring the quality of the environment to control environmental pollution is known as environmental monitoring. It is majorly used to prepare environmental impact assessments. It focuses on three major areas, namely, air, soil and water. Air quality monitoring is the assessment of air pollutants through air sampling. Soil monitoring is the collection and assessment of soil to ensure its quality and physical status. It is done by soil sampling, which includes monitoring of soil contamination, salinity and erosion. Water quality monitoring includes examining the water quality by collecting and analyzing water samples. A wide range of sampling methods including judgmental sampling and simple random sampling are used for this purpose. The objective of this book is to give a general view of the different areas of environmental monitoring. Most of the topics introduced herein cover new techniques and applications of environmental monitoring. The textbook is appropriate for those seeking detailed information in this area.

A brief description of the chapters is provided below for further understanding:

Chapter - Introduction

Environmental monitoring refers to the activities and processes which are used to characterize and monitor the quality of the environment. It is used for the preparation of environmental impact assessments. This chapter has been carefully written to provide an easy understanding of the various aspects of environmental monitoring.

Chapter - Devices, Methods and Techniques

Some of the methods and techniques that are used in environmental monitoring are carbon profiling, participatory monitoring and atmospheric dispersion modeling. Environmental monitoring also makes use of devices such as barometer and rain gauge. The diverse applications of these devices, methods and techniques in environmental monitoring have been thoroughly discussed in this chapter.

Chapter - Air Quality Monitoring

The quality of air is monitored in terms of the degree by which the air is clean and free from pollutants like smog, dust, smoke and other impurities of the air. Air quality index, air pollution forecasting and monitoring are some of the aspects that fall under its domain. The topics elaborated in this chapter will help in gaining a better perspective about air quality monitoring.

Chapter - Soil Monitoring

Soil is a mixture of a number of organic matters, gases, minerals, liquids and organisms that support life. Soil monitoring includes the collection and analysis of soil, its quality and constituents. Soil erosion and soil pollution are a few of the focus areas of soil monitoring. This chapter discusses these aspects related to soil monitoring in detail.

Chapter - Water Quality Monitoring

The chemical, biological and physical, and radiological characteristics of water can be termed as water quality. Its monitoring refers to the assessment of the qualitative information of these characteristics of water. This chapter closely examines these key concepts of water quality monitoring to provide an extensive understanding of the subject.

Clara Simon

1

Introduction

Environmental monitoring refers to the activities and processes which are used to characterize and monitor the quality of the environment. It is used for the preparation of environmental impact assessments. This chapter has been carefully written to provide an easy understanding of the various aspects of environmental monitoring.

ENVIRONMENT

The term 'environment' is widely used and has a broad range of definitions, meanings and interpretations. What does the term 'environment' mean? In popular usage, for some people, the term 'environment' means, simply, 'nature': in other words, the natural landscape together with all of its non-human features, characteristics and processes. To those people, the environment is often closely related to notions of wilderness and of pristine landscapes that have not been influenced - or, at least, that have been imperceptibly influenced - by human activities. However, for other people, the term 'environment' includes human elements to some extent. Many people would regard agricultural and pastoral landscapes as being part of the environment, whilst others are yet more inclusive and regard all elements of the earth's surface - including urban areas - as constituting the environment. Thus, in popular usage, the notion of the 'environment' is associated with diverse images and is bound up with various assumptions and beliefs that are often unspoken - yet may be strongly held. All of these usages, however, have a central underlying assumption: that the 'environment' exists in some kind of relation to humans. Hence the environment is, variously, the 'backdrop' to the unfolding narrative of human history, the habitats and resources that humans exploit, the 'hinterland' that surrounds human settlements, or the 'wilderness' that humans have not yet domesticated or dominated.

In its most literal sense, 'environment' simply means 'surroundings' (environs); hence the environment of an individual, object, element or system includes all of the other entities with which it is surrounded. However, in reality, individuals, objects, elements and systems rarely exist in isolation; instead, they tend to interact to varying extents with their surrounding entities. Therefore, it is not particularly helpful to conceptualise the 'environment' without including in that conceptualisation some notion of relationship. Individuals, objects, elements and systems influence - and are in turn influenced by - their surroundings. Indeed, the networks of relationships that exist between different entities may, in some cases, be extensive and highly complex. Thus the 'environment' may be regarded as a 'space' or a 'field' in which networks of relationships, interconnections and interactions between entities occur. To those who have studied the science of ecology, such a conceptualisation will be familiar, since ecologists are concerned with both the biotic (living) and abiotic (non-living) components of environmental systems - and especially with the interactions of those components. In fact, the term 'environment' is often used interchangeably with an ecological term 'ecosystem', which may be defined as a community of interacting organisms together with their physical surroundings. The notion of interrelationship is a central one in environmental science and management, since many environmental issues have occurred because one environmental system has been disturbed or degraded - either accidentally or deliberately - as a result of changes in another.

ENVIRONMENTAL MONITORING

Environmental monitoring involves the assessment of the quality of the environment in order to control the risk of pollution.

It is used as the basis of the production of environmental impact assessments. This information is then used to study environmental trends and to quantify the current state of the environment.

Assessments through environmental monitoring are subject to statistical analysis. Environmental monitoring utilizes specialized equipment and specific monitoring methods to get the most accurate recording of pollutant levels.

Environmental monitoring takes the form of air, soil, and water monitoring. Air monitors are used to identify the air quality and the levels of pollution. Soil monitoring requires the collection of soil samples to be analyzed and tested in a laboratory setting. This may be by way of grab sampling or composite sampling. Water monitoring may take the form of chemical, biological, radiological, or microbiological testing.

The sampling method used depends on the type of environment, the sampling material, and the future use of the data collected.

Air Monitoring

Air pollutants are known for their adverse effects on human health and ecosystems. Some of these pollutants also erode technical infrastructure and cultural monuments.

Emissions of nitrogen oxides and non-methane volatile organic compounds are the main causes of the formation of ground-level ozone, which has adverse effects on human health and ecosystems. The air pollutants indicator assesses pressures from specific pollutants on atmospheric air across individual countries, but also identifies pressures from particular national sectors like energy, transport, industrial processes, agriculture and waste management.

On the basis of this indicator, public authorities can adjust the national environmental policy by, for instance, revising emission standards and emission limit values, strengthening permitting of potentially polluting activities and improving the application of economic instruments. Information on pollutant emissions is necessary for the assessment of transboundary air pollution and for international cooperation to address this problem.

Water Monitoring

Renewable freshwater resources have major environmental and economic value. Their distribution varies widely among and within countries. Pressures on freshwater resources are exerted by overexploitation and by pollution. Relating resources abstraction to renewal of stocks is a central issue in sustainable freshwater resource management. If a significant share of a country's water comes from transboundary rivers, tensions between countries can arise, especially if water availability in the upstream country is greater than in the downstream one. Countries are quite interdependent with regard to water resources.

The Convention on the Protection and Use of Transboundary Watercourses and International Lakes requires that the Parties introduce sustainable water management, including an ecosystem approach and the rational and fair use of transboundary waters.

Biodiversity

Sustainable development depends on a sound environment, which in turn depends on ecosystem diversity. Protected areas, especially the full range of International Union for Conservation of Nature (IUCN) Protected Area Categories, are essential for conserving biodiversity and contributing to sustainable development.

The biodiversity indicator provides a means to measure the response to the degradation of ecosystems and the loss of biodiversity in a country. It demonstrates the extent to which areas important for conserving biodiversity, cultural heritage, scientific research, recreation, natural resource maintenance and other environmental values are protected from incompatible uses.

Waste Monitoring

Waste represents a considerable loss of resources in the form of materials and energy. The treatment and disposal of waste may cause environmental pollution and expose humans to harmful substances and infectious organisms. Waste generation is closely linked to the level of economic activity in a country, and reflects the society's production and consumption patterns. A reduction in the volume of waste generated is therefore an indication of an economy's move towards less material-intensive production and consumption patterns.

The main purpose of the waste indicator is to measure the pressure on the environment of the total amount of generated waste and waste by category. The waste intensity represents a driving force indicator and shows response to anthropogenic activities.

Remote Sensing

Remote sensing can play an important role in monitoring and reporting on environmental problems, in particular when the target of such observations is to assess pollutant effects at large space scales over long time periods — i.e., at the regional, continental or even global scale over the entire seasonal cycle for a number of years.

Remote sensing can provide complementary information to existing ground-based environmental monitoring systems. It can be used to meet the need for timely information and can provide synoptic cross-boundary information. Data and information obtained through Earth observation can be used within geographic information systems for overlay and comparison with other geo-referenced information.

Enterprise Monitoring

Enterprise environmental monitoring and reporting is the system of measures implemented and paid for by operators, the natural or legal persons exercising actual power over the technical functioning of the facility. Such a system includes continuous and periodic observations, the recording, storage and treatment of data relating to environmental protection and the reporting of the results to the management and all the employees of enterprises, the public authorities and the general public as sets of primary, calculated or aggregated data and general information.

Strengthening enterprise environmental monitoring and reporting will improve monitoring of enterprise compliance with environmental regulations. Increasing the quantity of environmental information produced by enterprises, improving the quality of this information and enhancing access to it by the general public will help to exert significant pressure on polluters to reduce their adverse environmental impact.

Environmental monitoring is critical to the protection of human health and the environment. As the human population continues to increase, as industrial development and energy use continues to expand, and despite advances in pollution control, the continued production of pollution remains inevitable. Thus the need for environmental monitoring is still as great as ever. Continued advances in the development, application, and automation of monitoring devices are needed to enhance the accuracy and cost-effectiveness of monitoring programs. Equally as important is the need to produce more scientists and engineers that have the knowledge and training required to successfully develop and operate monitoring devices and manage monitoring programs.

Environmental monitoring data is often used in risk assessment to create environmental regulations. However, the task of interpreting data is critical in making decisions about potential health risks and corrective actions. Risk assessment is a process that has been formalized to estimate the risk of adverse health effects caused by exposure to harmful chemicals and microorganisms. The goal is to express risk caused by exposure to a contaminant in terms of probability of illness or mortality. In this format such information can be better utilized by decision makers to determine the magnitude of the problem and weigh the costs and benefits of prevention or corrective action.

Types of Sensors for Environmental Monitoring

There are numerous types of environmental sensors depending on the specifications needed for monitoring. Sensitivity, size and speed are all factors to consider when choosing an environmental sensor. Sensors can be divided into categories depending on the type of sensor being used, and the environmental factors that an analyst is looking at.

The main types of sensors include:

- Trace metal sensors: Laser-induced breakdown spectroscopy (LIBS), nanoelectrode array and miniature chemical flow-probe sensor.
- Radioisotopic sensors: RadFET (Radiation field-effect transistor), cadmium

zinc telluride (CZT) detectors, low-energy pin diodes beta spectrometer, thermoluminescent dosimeter (TLD), isotope identification gamma detector, neutron generator for nuclear material detection, non-sandia radiation detectors.

- Volatile organic compound sensors: Evanescent fiber-optic chemical sensor, grating light reflection spectroscopy (GLRS), miniature chemical flow probe sensor, chemical sensor arrays, MicroChemLab (gas phase), gold nanoparticle chemiresistors, electrical impedance of tethered lipid bilayers on planar electrodes, MicroHound, hyperspectral imaging, chemiresistor array.
- Biological sensors: Fatty acid methyl esters (FAME) analyser, insulator-based dielectrophoresis, biological sensor arrays, μProLab, MicroChemLab (Liquid).

Sensors can also be split into types depending on environmental factors. These types of sensors include:

- Moisture sensors: Moisture sensors are essential for the measurement of volumetric water content in the soil towards the chimerical environment by several orders of magnitude.
- Pressure sensors/transducers: Pressure sensors operate on the basis of pressure applied. They are used in fields such as aviation, manufacturing biomedical measurements, auto mobile and hydraulic measurements.
- Tilt sensors: The tilt sensor relates the two different axes indicating a reference plane which may be in two different axes. This plays an important role in measuring tilt angles with reference to the Earth's ground plane. Tilt sensors are common in industry and in game controllers.
- Rain sensors: Rain sensors are a type of switching device which gets activated in the presence of rainfall. Rain sensors are used in areas such as as water conservation devices in irrigation systems, and in automobiles with windscreen wipers.
- Vibration sensors: Vibration sensors are important for displacement and acceleration with respect to the enduring impact on the environment. Displacement, linear velocity and acceleration are different factors which vibration sensors are measured according to.

2

Devices, Methods and Techniques

Some of the methods and techniques that are used in environmental monitoring are carbon profiling, participatory monitoring and atmospheric dispersion modeling. Environmental monitoring also makes use of devices such as barometer and rain gauge. The diverse applications of these devices, methods and techniques in environmental monitoring have been thoroughly discussed in this chapter.

BAROMETER

A barometer is a scientific instrument used to measure atmospheric pressure, also called barometric pressure. The atmosphere is the layers of air wrapped around the Earth. That air has a weight and presses against everything it touches as gravity pulls it to Earth. Barometers measure this pressure.

Atmospheric pressure is an indicator of weather. Changes in the atmosphere, including changes in air pressure, affect the weather. Meteorologists use barometers to predict short-term changes in the weather.

A rapid drop in atmospheric pressure means that a low-pressure system is arriving. Low pressure means that there isn't enough force, or pressure, to push clouds or storms away. Low-pressure systems are associated with cloudy, rainy, or windy weather. A rapid increase in atmospheric pressure pushes that cloudy and rainy weather out, clearing the skies and bringing in cool, dry air.

A barometer measures atmospheric pressure in units of measurement called atmospheres or bars. An atmosphere (atm) is a unit of measurement equal to the average air pressure at sea level at a temperature of 15 degrees Celsius (59 degrees Fahrenheit).

The number of atmospheres drops as altitude increases because the density of air is lower and exerts less pressure. As altitude decreases, the density of air increases, as does the number of atmospheres. Barometers have to be adjusted for changes in altitude in order to make accurate atmospheric pressure readings.

Working Method

A typical barometer consists of a glass tube with opened and closed ends. The tube filled with mercury is fixed upside down in a reservoir containing mercury and a ruler is placed alongside the tube. A vacuum is created at the top when the mercury level in the glass tube falls.

The barometer is operated by balancing mercury weight in the glass tube over the weight of atmospheric pressure. The mercury level in the glass tube increases, if the weight of mercury is less than the atmospheric pressure. The mercury level decreases if the weight of mercury is greater than the atmospheric pressure.

However, the mercury level is subjected to change constantly till the weight of the mercury in the tube is equal to the weight of the air above the reservoir. The rising and falling of mercury level can be measured using the ruler, and thus recording the changes with time.

Types of Barometers

Water-based Barometers

Goethe's device.

The concept that decreasing atmospheric pressure predicts stormy weather, postulated by Lucien Vidi, provides the theoretical basis for a weather prediction device called a “weather glass” or a “Goethe barometer” (named for Johann Wolfgang Von Goethe, the renowned German writer and polymath who developed a simple but effective weather ball barometer using the principles developed by Torricelli). The

French name, *le baromètre Liègeois*, is used by some English speakers. This name reflects the origins of many early weather glasses – the glass blowers of Liège, Belgium.

The weather ball barometer consists of a glass container with a sealed body, half filled with water. A narrow spout connects to the body below the water level and rises above the water level. The narrow spout is open to the atmosphere. When the air pressure is lower than it was at the time the body was sealed, the water level in the spout will rise above the water level in the body; when the air pressure is higher, the water level in the spout will drop below the water level in the body. A variation of this type of barometer can be easily made at home.

Mercury Barometers

A mercury barometer has a vertical glass tube closed at the top sitting in an open mercury-filled basin at the bottom. Mercury in the tube adjusts until the weight of the mercury column balances the atmospheric force exerted on the reservoir. High atmospheric pressure places more force on the reservoir, forcing mercury higher in the column. Low pressure allows the mercury to drop to a lower level in the column by lowering the force placed on the reservoir. Since higher temperature levels around the instrument will reduce the density of the mercury, the scale for reading the height of the mercury is adjusted to compensate for this effect. The tube has to be at least as long as the amount dipping in the mercury + head space + the maximum length of the column.

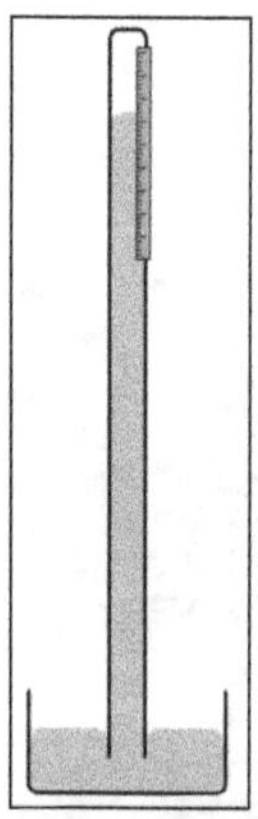

Schematic drawing of a simple mercury barometer with vertical mercury column and reservoir at base.

Torricelli documented that the height of the mercury in a barometer changed slightly each day and concluded that this was due to the changing pressure in the atmosphere. He wrote: "We live submerged at the bottom of an ocean of elementary air, which is known by incontestable experiments to have weight". Inspired by Torricelli, Otto von Guericke on 5 December 1660 found that air pressure was unusually low and predicted a storm, which occurred the next day.

Fortin barometer.

The mercury barometer's design gives rise to the expression of atmospheric pressure in inches or millimeters of mercury (mmHg). A torr was originally defined as 1 mmHg. The pressure is quoted as the level of the mercury's height in the vertical column. Typically, atmospheric pressure is measured between 26.5 inches (670 mm) and 31.5 inches (800 mm) of Hg. One atmosphere (1 atm) is equivalent to 29.92 inches (760 mm) of mercury.

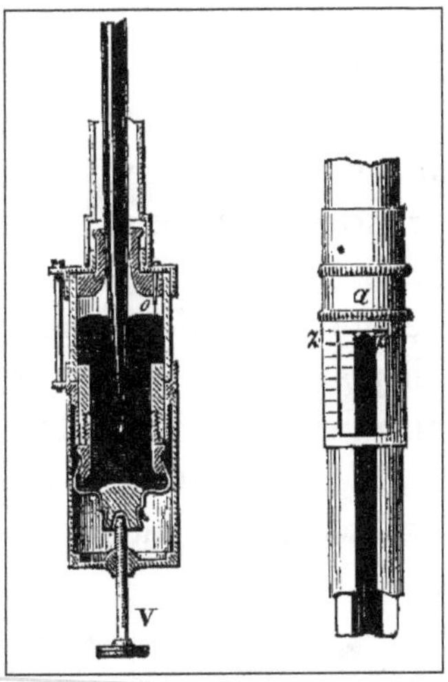

Reservoir of a Fortin barometer.

Design changes to make the instrument more sensitive, simpler to read, and easier to transport resulted in variations such as the basin, siphon, wheel, cistern, Fortin, multiple folded, stereometric, and balance barometers. *Fitzroy* barometers combine the standard mercury barometer with a thermometer, as well as a guide of how to interpret pressure changes. Fortin barometers use a variable displacement mercury cistern, usually constructed with a thumbscrew pressing on a leather diaphragm bottom (V in the diagram). This compensates for displacement of mercury in the column with varying pressure. To use a Fortin barometer, the level of mercury is set to zero by using the thumbscrew to make an ivory pointer (O in the diagram) just touch the surface of the mercury. The pressure is then read on the column by adjusting the vernier scale so that the mercury just touches the sightline at Z. Some models also employ a valve for closing the cistern, enabling the mercury column to be forced to the top of the column for transport. This prevents water-hammer damage to the column in transit.

On June 5, 2007, a European Union directive was enacted to restrict the sale of mercury, thus effectively ending the production of new mercury barometers in Europe.

Vacuum Pump Oil Barometer

Using vacuum pump oil as the working fluid in a barometer has led to the creation of the new "World's Tallest Barometer" in February 2013. The barometer at Portland State University (PSU) uses doubly distilled vacuum pump oil and has a nominal height of about 12.4 m for the oil column height; expected excursions are in the range of ±0.4 m over the course of a year. Vacuum pump oil has very low vapor pressure and it is available in a range of densities; the lowest density vacuum oil was chosen for the PSU barometer to maximize the oil column height.

Aneroid Barometers

An aneroid barometer is an instrument used for measuring pressure as a method that does not involve liquid. Invented in 1844 by French scientist Lucien Vidi, the aneroid barometer uses a small, flexible metal box called an aneroid cell (capsule), which is made from an alloy of beryllium and copper. The evacuated capsule (or usually several capsules, stacked to add up their movements) is prevented from collapsing by a strong spring. Small changes in external air pressure cause the cell to expand or contract. This expansion and contraction drives mechanical levers such that the tiny movements of the capsule are amplified and displayed on the face of the aneroid barometer. Many models include a manually set needle which is used to mark the current measurement so a change can be seen. This type of barometer is common in homes and in recreational boats. It is also used in meteorology, mostly in barographs and as a pressure instrument in radiosondes.

Old aneroid barometer.

Modern aneroid barometer.

Barographs

A barograph is a recording aneroid barometer where the changes in atmospheric pressure are recorded on a paper chart.

Principle and Working

The principle of the barograph is same as that of the aneroid baromoter. A metal box partially exhausted of air will undergo changes of shape as the outside pressure varies. These small movements of the box are transmitted by a system of levers to a recording arm that has at its extreme end either a scribe or a pen. A scribe records on smoked foil while a pen records on paper using ink, held in a nib. The recording material is mounted on a cylindrical drum which is rotated slowly by a clock. Commonly, the drum makes one revolution per day, per week, or per month and the rotation rate can often be selected by the user.

MEMS Barometers

Micro Electro Mechanical Systems (or MEMS) barometers are extremely small devices between 1 and 100 micrometres in size (0.001 to 0.1 mm). They are created via photolithography or photochemical machining. Typical applications include miniaturized weather stations, electronic barometers and altimeters.

A barometer can also be found in smartphones such as the Samsung Galaxy Nexus, Samsung Galaxy S3-S6, Motorola Xoom, Apple iPhone 6 smartphones, and Timex Expedition WS4 smartwatch, based on MEMS and piezoresistive pressure-sensing technologies. Inclusion of barometers on smartphones was originally intended to provide a faster GPS lock. However, third party researchers were unable to confirm additional GPS accuracy or lock speed due to barometric readings. The researchers suggest that the inclusion of barometers in smartphones may provide a solution for determining a user's elevation, but also suggest that several pitfalls must first be overcome.

More Unusual Barometers

Barometric chart mode with weather forecast function.

There are many other more unusual types of barometer. From variations on the storm barometer, such as the Collins Patent table Barometer, to more traditional-looking designs such as Hooke's Otheometer and the Ross Sympiesometer. Some, such as the Shark Oil barometer, work only in a certain temperature range, achieved in warmer climates.

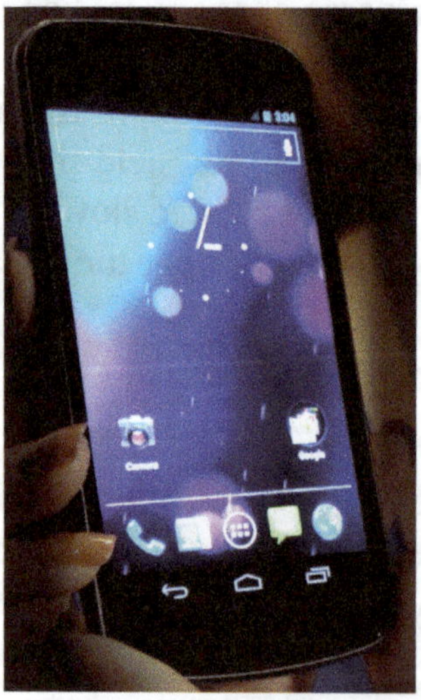

The Galaxy Nexus has a built-in barometer.

Compensations

Temperature

The density of mercury will change with increase or decrease in temperature, so a reading must be adjusted for the temperature of the instrument. For this purpose a mercury thermometer is usually mounted on the instrument. Temperature compensation of an aneroid barometer is accomplished by including a bi-metal element in the mechanical linkages. Aneroid barometers sold for domestic use typically have no compensation under the assumption that they will be used within a controlled room temperature range.

Altitude

A digital barometer with altimeter setting (for correction) displayed.

As the air pressure decreases at altitudes above sea level (and increases below sea level) the uncorrected reading of the barometer will depend on its location. The reading is then adjusted to an equivalent sea-level pressure for purposes of reporting. For example, if a barometer located at sea level and under fair weather conditions is moved to an altitude of 1,000 feet (305 m), about 1 inch of mercury (~35 hPa) must be added on to the reading. The barometer readings at the two locations should be the same if there are negligible changes in time, horizontal distance, and temperature. If this were not done, there would be a false indication of an approaching storm at the higher elevation.

Aneroid barometers have a mechanical adjustment that allows the equivalent sea level pressure to be read directly and without further adjustment if the instrument is not moved to a different altitude. Setting an aneroid barometer is similar to resetting an analog clock that is not at the correct time. Its dial is rotated so that the current atmospheric pressure from a known accurate and nearby barometer (such as the local weather station) is displayed. No calculation is needed, as the source barometer reading has already been converted to equivalent sea-level pressure, and this is transferred to the barometer being set—regardless of its altitude. Though somewhat rare, a few aneroid barometers intended for monitoring the weather are calibrated to manually adjust for altitude. In this case, knowing *either* the altitude or the current atmospheric pressure would be sufficient for future accurate readings.

The table shows examples for three locations in the city of San Francisco, California. Note the corrected barometer readings are identical, and based on equivalent sea-level pressure. (Assume a temperature of 15 °C).

Location	Altitude (feet)	Uncorrected P_{atm} (inches Hg)	Corrected P_{atm} (inches Hg)	Altitude (metres)	Uncorrected P_{atm} (hPa)	Corrected P_{atm} (hPa)
City Marina	Sea Level (0)	29.92	29.92	0 m	1013 hPa	1013 hPa
Nob Hill	348	29.55	29.92	106 m	1001 hPa	1013 hPa
Mt. Davidson	928	28.94	29.92	283 m	980 hPa	1013 hPa

When atmospheric pressure is measured by a barometer, the pressure is also referred to as the "barometric pressure". Assume a barometer with a cross-sectional area *A*, a height *h*, filled with mercury from the bottom at Point B to the top at Point C. The pressure at the bottom of the barometer, Point B, is equal to the atmospheric pressure. The pressure at the very top, Point C, can be taken as zero because there is only mercury vapor above this point and its pressure is very low relative to the atmospheric pressure. Therefore, one can find the atmospheric pressure using the barometer and this equation:

$$P_{atm} = \rho g h$$

where ρ is the density of mercury, g is the gravitational acceleration, and h is the height of the mercury column above the free surface area. The physical dimensions (length of tube and cross-sectional area of the tube) of the barometer itself have no effect on the height of the fluid column in the tube.

In thermodynamic calculations, a commonly used pressure unit is the "standard atmosphere". This is the pressure resulting from a column of mercury of 760 mm in height at 0 °C. For the density of mercury, use $\rho_{Hg} = 13{,}595$ kg/m³ and for gravitational acceleration use $g = 9.807$ m/s².

If water were used (instead of mercury) to meet the standard atmospheric pressure, a water column of roughly 10.3 m (33.8 ft) would be needed.

Standard atmospheric pressure as a function of elevation:

1 torr = 133.3 Pa = 0.03937 In Hg

P_{atm}	Altitude	P_{atm}	Altitude
101.325 kPa	Sea Level (0m)	29.92 In Hg	Sea Level (0 ft)
97.71 kPa	305 m	28.86 In Hg	1,000 ft
94.21 kPa	610 m	27.82 In Hg	2,000 ft
89.88 kPa	1,000 m	26.55 In Hg	3,281 ft
84.31 kPa	1,524 m	24.90 In Hg	5,000 ft
79.50 kPa	2,000 m	23.48 In Hg	6,562 ft
69.68 kPa	3,048 m	20.58 In Hg	10,000 ft
54.05 kPa	5,000 m	15.96 In Hg	16,404 ft
46.56 kPa	6,096 m	13.75 In Hg	20,000 ft

37.65 kPa	7,620 m	11.12 In Hg	25,000 ft
32.77 kPa	8,848 m*	9.68 In Hg	29,029 ft*
26.44 kPa	10,000 m	7.81 In Hg	32,808 ft
11.65 kPa	15,240 m	3.44 In Hg	50,000 ft
5.53 kPa	20,000 m	1.63 In Hg	65,617 ft

PARTICULATE MATTER SAMPLER

A particulate matter sampler is an instrument for measuring the properties (such as mass concentration or chemical composition) of particulates in the ambient air.

Types

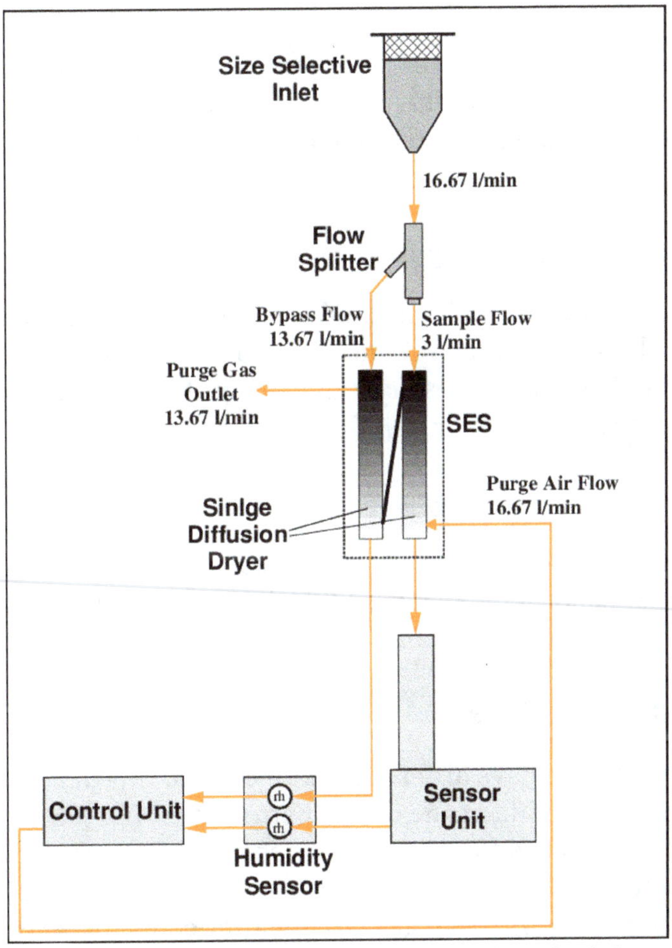

Schematic diagram of a tapered element oscillating microbalance.

Two different types of particulate matter samplers exist that measure particulate mass concentration: Manual samplers and automated samplers.

Manual

Manual samplers draw a known volume of air through a filter. The filter is weighed on an analytical balance before and after sampling, and the difference in weight divided by the volume of air pulled through the filter gives the mass concentration of the particulate.

Automated

Automated samplers do the weighing in the field. There are two types of automated samplers in common usage: samplers that use a beta gauge for mass measurement and samplers that use a tapered element oscillating microbalance (TEOM) for mass measurement.

Beta Gauge

Beta gauge particulate samplers have an appearance that is similar to a reel to reel tape recorder. Air is pulled through a filter tape to accumulate a sample, the mass of the tape before and after sampling is determined by advancing the tape spot into the beta attenuation cell.

Tapered Element Oscillating Microbalance

The tapered element oscillating microbalance (TEOM) particulate sampler operates by drawing air through a filter attached at the tip of a glass tube. An electrical circuit places the tube into oscillation, and the resonant frequency of the tube is proportional to the square root of the mass on the filter.

Noise based Particle Counters

Recently, microphone based instruments have been devised that monitor noise levels in specific frequency bands to predict local PNC levels. Prototypes of such instruments have been tested in Europe and in Bangalore.

Use of Inertial Separators

Particles of different sizes have different health effects. Inertial separators are used to eliminate particles outside of the desired size range. If a gas stream containing particles of different sizes is forced to turn a sharp corner, the inertia of the large particles causes them to separate from the gas stream lines. The larger particles can be collected and removed from the gas stream after collisions with the walls of the vessel.

The two common types of inertial separators are cyclones, which spin the gas stream, causing collisions of the heavier particles with the outside of the cyclone wall, and impactors, where the gas particle stream is directed at a greased metal plate and turned at the last moment, causing the larger particles to stick to the greased plate.

Modern particulate samplers use a volumetric flow control system that pulls air through the particle separator at the velocity required to achieve the desired cutpoint.

For air pollution applications, the definition of "particulate" does not include uncombined water, and water from a particulate sample must be removed before it is weighed. This can be done either by heating the sample to evaporate the water or by placing the sample in a low humidity environment before weighing.

POLAR ORGANIC CHEMICAL INTEGRATIVE SAMPLER

A polar organic chemical integrative sampler (POCIS) is a passive sampling device which allows for the in situ collection of a time-integrated average of hydrophilic organic contaminants developed by researchers with the United States Geological Survey in Columbia, Missouri. POCIS provides a means for estimating the toxicological significance of waterborne contaminants. The POCIS sampler mimics the respiratory exposure of organisms living in the aquatic environment and can provide an understanding of bioavailable contaminants present in the system. POCIS can be deployed in a wide range of aquatic environments and is commonly used to assist in environmental monitoring studies.

The first passive sampling devices were developed in the 1970s to determine concentrations of contaminants in the air. In 1980 this technology was first adapted for the monitoring of organic contaminants in water. The initial type of passive sampler developed for aquatic monitoring purposes was the semipermeable membrane device (SMPD). SPMD samplers are most effective at absorbing hydrophobic pollutants with a log octanol-water partitioning coefficient (K_{ow}) ranging from 4-8. As the global emission of bioconcentratable persistent organic pollutants (POPs) was shown to result in adverse ecological effects, industry developed a wide range of increasing water-soluble, polar hydrophilic organic compounds (HpOCs) to replace them. These compounds generally have lower bioconcentration factors. However, there is evidence that large fluxes of these HpOCs into aquatic environments may be responsible for a number of adverse effects to aquatic organisms, such as altered behavior, neurotoxicity, endocrine disruption, and impaired reproduction. In the late 1990s research was underway to develop a new passive sampler in order to monitor HpOCs with a log Kow value of less than 3. In 1999 the POCIS sampler was under development at the University of Missouri-Columbia. It gathered more support in the early 2000s as concern increased regarding the effects of pharmaceutical and personal care products in surface waters.

The United States Geological Survey (USGS) has been heavily involved in the development of passive samplers and has articles in their database regarding the development of POCIS as early as 2000. The USGS Columbia Environmental Research Center (CERC) is a self-proclaimed international leader in the field of passive sampling. There have been recent efforts by the USGS to connect people who have an interest in passive sampling. An international workshop and symposium on passive sampling was held by the USGS in 2013 to connect developers, policy makers and end users in order to discuss ways of monitoring environmental pollution.

Fundamentals

The POCIS device was developed and patented by Jimmie D. Petty, James N. Huckins, and David A. Alvarez, of the Columbia Environmental Research Center. Integrative passive samplers are an effective way to monitor the concentration of organic contaminants in aquatic systems over time. Most aquatic monitoring programs rely on collecting individual samples, often called grab samples, at a specific time. The grab sampling method is associated with many disadvantages that can be resolved by passive sampling techniques. When contaminants are present in trace amounts, grab sampling may require the collection of large volumes of water. Also, lab analysis of the sample can only provide a snapshot of contaminant levels at the time of collection. This approach therefore has drawbacks when monitoring in environments where water contamination varies over time and episodic contamination events occur. Passive sampling techniques have been able to provide a time-integrated sample of water contamination with low detection limits and in situ extraction of analytes.

POCIS Set-up

The POCIS sampler consists of an array of sampling disks mounted on a support rod. Each disk consists of a solid sorbent sandwiched between two polyethersoulfone (PES) microporous membranes which are then compressed between two stainless steel rings which expose a sampling area. A standard POCIS disk consists of a sampling surface area to sorbent mass ratio of approximately 180 cm^2g. Because the amount of chemical sampled is directly related to the sample surface area, it is sometimes necessary to combine extracts from multiple POCIS disks into one sample. Stainless steel rings, or other rigid inert material, are essential to prevent sorbent loss as the PES membranes are not able to be heat sealed. The POCIS array is then inserted and deployed within a protective canister. This canister is usually made of stainless steel or PVC and works to deflect debris that may displace the POCIS array during its deployment.

The PES membrane acts as a semipermeable barrier between the sorbent and surrounding aquatic environment. It allows dissolved contaminants to pass through the sorbent while selectively excluding any particles larger than 100 nm. The membrane resists biofouling because the polyethersulphone used in the design is less prone than other materials. The POCIS is versatile in that the sorbents can be changed to target different classes of contaminants. However, only two sorbent classes are considered as standards of all POCIS deployments to date.

Theory and Modeling

Each POCIS disk will sample a certain volume of water per day. The volume of water sampled varies from chemical to chemical and is dependent on the physical and chemical properties of the compound as well as the duration of sampling. The sampling rate of POCIS can vary with changes in the water flow, turbulence, temperature, and the buildup of solids on the sampler's surface. The accumulation of contaminants into a POCIS device is the result of three successive process occurring at the same time. First, the contaminants have to diffuse across the water boundary layer. The thickness of this layer is dependent on water flow and turbulence around the sampler and can significantly alter sampling rates. Second, the contaminant must transport across the membrane either through the water-filled pores or through the membrane itself. Finally, contaminants transfer from the membrane into the sorbent material mainly through adsorption. These last two steps make the modeling, understanding, and prediction of accumulation by a POCIS device challenging. To date, a limited number of chemical sampling rates have been determined.

Accumulation of chemicals by a POCIS device generally follows first order kinetics. The kinetics are characterized by an initial integrative phase, followed by an equilibrium partitioning phase. During the integrative phase of uptake, a passive sampling device accumulates residues linearly relative to time, assuming constant exposure concentrations. Based on current results, the POCIS sampler remains in a linear phase for at least 30 days, and has been observed up to 56 days. Therefore, both laboratory and field data justify the use of a linear uptake model for the calculation of sample rates. In order to estimate the ambient water concentration of contaminants sampled by a POCIS device, there must be available calibration data applicable for in situ conditions regarding the target compound. Currently, this information is limited.

Applicability

POCIS can be deployed in a wide range of aquatic environments including stagnant pools, rivers, springs, estuarine systems, and wastewater streams. However, there has been little research into the use of POCIS in strictly marine environments. Prior to deployment of a POCIS device, it is essential to select a study site that will maximize the effectiveness of the sampler. Selecting an area that is shaded will help prevent light sensitive chemicals from being degrading. The site should also allow the sampler to be submerged in the water without being buried in the sediment. It is ideal to place the sampler in moving water in order to increase sampling rates, however, areas with an extremely turbulent water flow should be avoided as to prevent damage to the POCIS device. Passive samplers are very vulnerable to vandalism and it is therefore important to secure the sampler in areas that are not easily visible and that are away from areas frequently used by people.

POCIS samplers can be deployed for a period of time ranging from weeks to months. The shortest deployment lengths are typically 7 days but average 2–3 months. It is important to have a long enough deployment period to allow for adequate detection of contaminants at ambient environmental concentrations. Often, the two different types of POCIS devices will be deployed together in order to provide the greatest understanding of contamination. It is also important to deploy enough POCIS devices to ensure a large enough sample of contaminant is recovered for chemical analysis. An estimate or the number of samplers needed at a given site can be determined by the following equation.

$$R_s \text{ x } t \text{ x } n \text{ x } C_c \text{ x } P_r \text{ x } E_t > MQL \text{ x } V_i$$

where:

- C_c is the predicted environmental concentration of the contaminant.
- t is the deployment time in days.
- R_s is sampling rate in liters of water extracted by the passive sampler per day(L/day).
- P_r is the overall method recovery for the analyte (expressed as a factor of one; ::therefore 0.9 is used for 90 percent recovery).
- n is the number of passive samplers combined into a single sample.
- E_t is the fraction of the total sample extract which is injected into the ::instrument for quantification.
- MQL is the method quantification limit.
- V_i is the volume of standard injection (commonly 1 μL).

Relevant Contaminants

Any compound with a log Kow of less than or equal to 3 can concentrate in a POCIS sampler. Applicable classes of contaminants measured by POCIS are pharmaceuticals, household and industrial products, hormones, herbicides, and polar pesticides. Currently, there are two POCIS configurations that are targeted for different classes of contaminants. A general POCIS design contains a sorbent that is used to collect pesticides, natural as well as synthetic hormones, and wastewater related chemicals. The pharmaceutical POCIS configuration contains a sorbent that is designed to specifically target classes of pharmaceuticals.

Applicable contaminants that concentrate in a POCIS device. Not to be considered a complete list.

Chemical Class	Examples
Pharmaceuticals	Acetaminophen, azithromycin, carbamazepine, ibuprofen, propranolol, sulfa drugs, tetracycline antibiotics
Household and industrial products	Alkyl phenols, benzophenone, caffeine, DEET, fire retardants, indole, triclosan
Hormones	17β-estradiol, 17α-ethynlestradiol, estrone, estriol
Herbicides	Atrazine, cyanazine, hydroxyatrazine, tertbutylazine
Polar pesticides	Alachlor, chlorppyrifos, diainon, dichlorvos, diuron, isoproturon, metolachlor
Other	Urobilin

POCIS Processing

Before the POCIS is constructed, all the hardware as well as the sorbents and membrane must be thoroughly cleaned so that any potential interference is removed. During and after sampling the only cleaning necessary is the removal of any sediment that has adhered to the surface of the sampler. After assembly, and prior to deployment, the samplers are stored in frozen airtight containers to avoid any contamination. The samplers should be kept in airtight containers during transportation both to and from the sampling site so that airborne contaminants do not contaminate the sampler. It is ideal to keep the samplers cold while transporting them in order to preserve the integrity of the samples.

After the POCIS is retrieved from the field, the membrane is gently cleaned to reduce the possibility of any contamination to the sorbent. The sorbent is placed into a chromatography column so that the chemicals that samples can be recovered using an organic solvent. The solvent used is specifically chosen based on the type of sorbent and chemicals sampled. The sample can go through further processing such as cleanup or fractionation depending on the desired use of the sample.

Data Analysis

After the sample has been processed, the extract can be analysed using a variety of data analysis techniques. The chemical analysis and analytical instrumentation used depends on the goal of the study. Many analyses require multiple samples, although in some cases a single POCIS sample can be used for multiple analyses.

It is vital to use quality control (QC) procedures when using passive samplers. It is common practice for 10% to 50% of the total number of samples to be used for QC purposes. The number of QC samples depends on the study objectives. The QC samples are used to address issues such as sample contamination and analyte recovery. The types of QC samples commonly used include; reagent blanks, field blanks, matrix spikes, and procedural spikes.

A large number of studies have been performed in which POCIS data was combined with bioassays to measure biological endpoints. Testing POCIS extracts in biological assays is useful as a POCIS device samples over its entire deployment period, and biologically active compounds can be effectively monitored. It can also be argued that the use of POCIS is a more relevant from an ecotoxicological perspective as the use of a passive sampler mimics the uptake of compounds by organisms. Another strength in using bioassays to test environmental samples is that they can provide an integrative measure of the toxic potential of a group of chemical compounds, rather than a single contaminant.

Other Passive Samplers

There are many types of passive samplers used that specialize in absorbing different classes of aquatic contaminants found in the environment. Chemcatcher and SMPD are two types of passive samplers that are also commonly used. Monitoring programs use SMPDs to measure to hydrophobic organic contaminants. SPMDs are designed to mimic the bioconcentration of contaminants in fatty tissues. Contaminants applicable to the use of an SPMD include, but are not limited to, polychlorinated biphenyls (PCBs), polycyclic aromatic hydrocarbons (PAHs), organochlorine pesticides, dioxins, and furans.

The SPMD consist of a thin-walled, nonporous, polyethylene membrane tube that is filled with high molecular weight lipid. These tubes are approximately 90 cm long and wrap around the inside of a stainless steel deployment canister. SMPDs are efficient at absorbing pollutants with a log Kow of 4-8. This slightly overlaps with the range of contaminants absorbed by POCIS. Because of this, SMPDs and POCIS devices are often used together in monitoring studies to achieve a more representative understanding of contamination.

ENVIRONMENTAL INDICATOR

Environmental indicators are simple measures that tell us what is happening in the environment. Since the environment is very complex, indicators provide a more practical and economical way to track the state of the environment than if we attempted to record every possible variable in the environment. For example, concentrations of ozone depleting substances (ODS) in the atmosphere, tracked over time, is a good indicator with respect to the environmental issue of stratospheric ozone depletion.

Environmental indicators have been defined in different ways but common themes exist.

"An environmental indicator is a numerical value that helps provide insight into the state of the environment or human health. Indicators are developed based on quantitative measurements or statistics of environmental condition that are tracked over time. Environmental indicators can be developed and used at a wide variety of geographic scales, from local to regional to national levels."

"A parameter or a value derived from parameters that describe the state of the environment and its impact on human beings, ecosystems and materials, the pressures on the environment, the driving forces and the responses steering that system. An indicator has gone through a selection and/or aggregation process to enable it to steer action."

Environmental indicator criteria and frameworks have been used to help in their selection and presentation.

It can be considered, for example, that there are major subsets of environmental indicators in-line with the Pressure-State-Response model developed by the OECD. One subset of environmental indicators is the collection of ecological indicators which can include physical, biological and chemical measures such as atmospheric temperature, the concentration of ozone in the stratosphere or the number of breeding bird pairs in an area. These are also referred to as "state" indicators as their focus is on the state of the environment or conditions in the environment. A second subset is the collection of indicators that measure human activities or anthropogenic pressures, such as greenhouse gas emissions. These are also referred to as "pressure" indicators. Finally, there are indicators, such as the number of people serviced by sewage treatment, which track societal responses to environmental issues.

Environmental indicators, in turn, should be considered as a subset of sustainable development indicators which are meant to track the overall sustainability of a society with respect to its environmental, social and economic integrity and health.

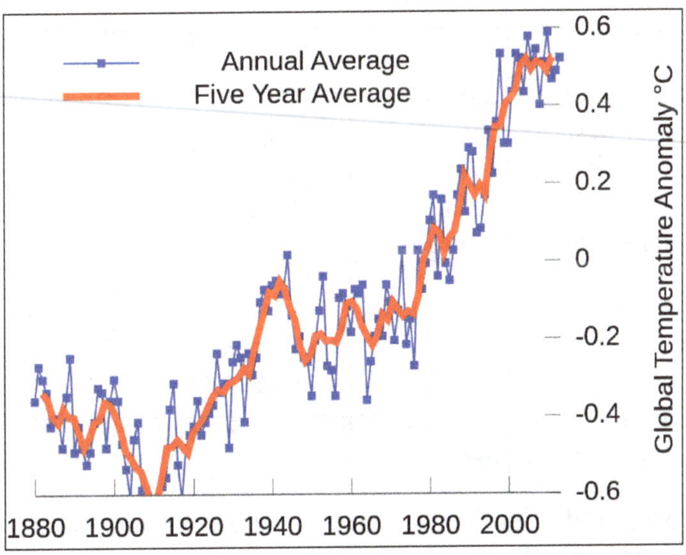

An example of an environmental indicator: Trend in global temperature anomalies of the last 150 years as an indicator of climate change.

A common framework spearheaded by the European Environment Agency is the "DPSIR" or "drivers, pressures, state, impact, response" framework. Drivers and pressures are indicators of the human activities and resulting pressures on the environment in the form of pollution or land-use change, for example. State and impact indicators are the resulting conditions in the environment and the implications for the health of ecosystems and humans. The response indicators measure the reaction of human society to the environmental issue. Criteria tend to focus on three key areas – scientific credibility, policy/social relevance and practical monitoring and data requirements.

Environmental indicators are used by governments, non-government organizations, community groups and research institutions to see if environmental objectives are being met, to communicate the state of the environment to the general public and decision makers and as a diagnostic tool through detecting trends in the environment.

Environmental indicators can be measured and reported at different scales. For example, a town may track air quality along with water quality and count the number of rare species of birds to estimate the health of the environment in their area. Indicators are developed for specific ecosystems, such as the Great-Lakes in North America. National governments use environmental indicators to show status and trends with respect to environmental issues of importance to their citizens.

Use in Assessment

Some have attempted to monitor and assess the state of the planet using indicators.

Environmentally, the world is in an overshoot mode. If we use environmental indicators to evaluate our situation, then the global decline of the economy's natural support systems - the environmental decline that will lead to economic decline and social collapse- is well underway.

Environmental indicators are also used by companies in the framework of an Environmental management system. The EU Eco-Management and Audit Scheme provides core indicators or Performance Indicator (KPIs) with which registered organizations can measure their environmental performance and monitor their continual environmental improvement against set targets.

The types of indicators selected or developed should be partially based on who will be using the information from the indicators. There are generally three possible audiences to consider, each with different information needs. These audiences are: 1) technical experts and science advisors, 2) policy-makers, decision makers and resource managers, and 3) general public and media.

The technical experts and scientists will be interested in detailed and complex indicators. These indicators should have scientific validity, sensitivity, responsiveness and have data available on past conditions. The audience that includes policy-makers and resource managers will be concerned with using indicators that are directly related to evaluating policies and objectives. They require their indicators to be sensitive, responsive and have historical data available like the technical audience, but they are also looking for indicators that are cost-effective and have meaning for public awareness. Finally, the general public responds to indicators that have clear and simple messages and are meaningful to them, such as the UV index and the air quality index.

Indicator Systems and Communicating Them

Individual indicators are designed to translate complex information in a concise and easily understood manner in order to represent a particular phenomenon (e.g. ambient air quality). In contrast, indicator systems (or collections of indicators), when seen as a whole are meant to provide an assessment of the full environment domain or a major subset of it (e.g. forests).

Some indicator systems have evolved to include many indicators and require a certain level of knowledge and expertise in various disciplines to fully grasp. A number of methods have been devised in the recent past to boil down this information and allow for rapid consumption by those who do not have the time or the expertise to analyse the full set of indicators. In general these methods can be categorized as *numerical aggregation* (e.g. indices), *short selections of indicators* (e.g. core set or headline indicators), *short visual assessments* (e.g. arrows, traffic signals), and *compelling presentations* (e.g. maps or the dashboard of sustainability). Many prominent environmental indicator systems have adjusted their indicator systems to include or report solely on a limited "indicator set" (e.g. the OECD's "Key Environmental Indicators" and the "Canadian Environmental Sustainability Indicators").

CARBON PROFILING

Carbon profiling is a mathematical process that calculates how much carbon dioxide is put into the atmosphere per m^2 of space in a building over one year.

The analysis has two parts, which are then added together to produce an overall figure which is termed the 'Carbon Profile':

- Operational carbon emissions.
- Embodied carbon emissions.

Embodied Carbon Emissions

Embodied carbon emissions relate to the amount of carbon dioxide emitted into the atmosphere from creating and maintaining the materials that form a building, e.g. the carbon dioxide released from the baking of bricks or smelting of iron. These emissions can also be considered to be Upfront Carbon Emissions, or UCE. "Embodied carbon refers to the carbon footprint associated with building materials, from cradle to grave," and can be quantified as a part of environmental impact using Life Cycle Assessment (LCA).

In the Carbon Profiling Model these emissions are measured as Embodied Carbon Efficiency (ECE), measured as kg of CO_2/m^2/year.

As of 2018, "Embodied carbon is responsible 11% of global GHG emissions and 28% of global building sector emissions. Embodied carbon will be responsible for almost half of total new construction emissions between now and 2050." Zero-carbon architecture (similar to zero-energy building), incorporates design techniques that maximize embodied carbon.

Occupational Carbon Emissions

Occupational carbon emissions relate to the amount of carbon dioxide emitted into the atmosphere from the direct use of energy to run the building e.g. the heating or electricity used by the building over the year. In the Carbon Profiling Model these emissions are measured in BER's (Building Emission Rate) in kg of $CO2/m^2$/year.

The BER is a United Kingdom government accepted unit of measurement that comes from an approved calculation process called sBEM (Simplified Building Emission Model).

The purpose of Carbon Profiling is to provide a method of analyzing and comparing both operational and embodied carbon emissions at the same time. With this information it is then possible to allocate a project's resources in such a way to minimize the total amount of Carbon Dioxide emitted into the atmosphere through the use of a given piece of space.

A secondary benefit is that having quantified the Carbon Profiling of different buildings it is then possible to make comparisons and rank buildings in term of their performance. This allows investors and occupiers to identify which building are good and bad carbon investments.

Simon Sturgis and Gareth Roberts of Sturgis Associates in the United Kingdom originally developed 'Carbon Profiling' in December 2007.

ATMOSPHERIC DISPERSION MODELING

Atmospheric dispersion modeling is the mathematical simulation of how air pollutants disperse in the ambient atmosphere. It is performed with computer programs that include algorithms to solve the mathematical equations that govern the pollutant dispersion. The dispersion models are used to estimate the downwind ambient concentration of air pollutants or toxins emitted from sources such as industrial plants, vehicular traffic or accidental chemical releases. They can also be used to predict future concentrations under specific scenarios (i.e. changes in emission sources). Therefore, they are the dominant type of model used in air quality policy making. They are most useful for pollutants that are dispersed over large distances and that may react in the atmosphere. For pollutants that have a very high spatio-temporal variability (i.e. have very steep distance to source decay such as black carbon) and for epidemiological studies statistical land-use regression models are also used.

Industrial air pollution source.

Dispersion models are important to governmental agencies tasked with protecting and managing the ambient air quality. The models are typically employed to determine whether existing or proposed new industrial facilities are or will be in compliance with the National Ambient Air Quality Standards (NAAQS) in the United States and other nations. The models also serve to assist in the design of effective control strategies to reduce emissions of harmful air pollutants. During the late 1960s, the Air Pollution Control Office of the U.S. EPA initiated research projects that would lead to the development of models for the use by urban and transportation planners. A major and significant application of a roadway dispersion model that resulted from such research was applied to the Spadina Expressway of Canada in 1971.

Air dispersion models are also used by public safety responders and emergency management personnel for emergency planning of accidental chemical releases. Models are used to determine the consequences of accidental releases of hazardous or toxic materials, Accidental releases may result in fires, spills or explosions that involve hazardous materials, such as chemicals or radionuclides. The results of dispersion modeling, using worst case accidental release source terms and meteorological conditions, can provide an estimate of location impacted areas, ambient concentrations, and be used to determine protective actions appropriate in the event a release occurs. Appropriate protective actions may include evacuation or shelter in place for persons in the downwind direction. At industrial facilities, this type of consequence assessment or emergency planning is required under the Clean Air Act (United States) (CAA) codified in Part 68 of Title 40 of the Code of Federal Regulations.

The dispersion models vary depending on the mathematics used to develop the model, but all require the input of data that may include:

- Meteorological conditions such as wind speed and direction, the amount of atmospheric turbulence (as characterized by what is called the "stability class"), the ambient air temperature, the height to the bottom of any inversion aloft that may be present, cloud cover and solar radiation.
- Source term (the concentration or quantity of toxins in emission or accidental release source terms) and temperature of the material.
- Emissions or release parameters such as source location and height, type of source (i.e., fire, pool or vent stack)and exit velocity, exit temperature and mass flow rate or release rate.
- Terrain elevations at the source location and at the receptor location(s), such as nearby homes, schools, businesses and hospitals.
- The location, height and width of any obstructions (such as buildings or other structures) in the path of the emitted gaseous plume, surface roughness or the use of a more generic parameter "rural" or "city" terrain.

Many of the modern, advanced dispersion modeling programs include a pre-processor module for the input of meteorological and other data, and many also include a post-processor module for graphing the output data and/or plotting the area impacted by the air pollutants on maps. The plots of areas impacted may also include isopleths showing areas of minimal to high concentrations that define areas of the highest health risk. The isopleths plots are useful in determining protective actions for the public and responders.

The atmospheric dispersion models are also known as atmospheric diffusion models, air dispersion models, air quality models, and air pollution dispersion models.

Atmospheric Layers

Discussion of the layers in the Earth's atmosphere is needed to understand where airborne pollutants disperse in the atmosphere. The layer closest to the Earth's surface is known as the *troposphere*. It extends from sea-level to a height of about 18 km and contains about 80 percent of the mass of the overall atmosphere. The *stratosphere* is the next layer and extends from 18 km to about 50 km. The third layer is the *mesosphere* which extends from 50 km to about 80 km. There are other layers above 80 km, but they are insignificant with respect to atmospheric dispersion modeling.

The lowest part of the troposphere is called the *atmospheric boundary layer (ABL)* or the *planetary boundary layer (PBL)* . The air temperature of the atmosphere decreases with increasing altitude until it reaches what is called an *inversion layer* (where the temperature increases with increasing altitude) that caps the Convective Boundary Layer, typically to about 1.5 to 2.0 km in height. The upper part of the troposphere (i.e., above the inversion layer) is called the *free troposphere* and it extends up to the tropopause (the boundary in the Earth's atmosphere between the troposphere and the stratosphere). In tropical and mid-latitudes during daytime, the Free convective layer can comprise the entire troposphere, which is up to 10 km to 18 km in the Intertropical convergence zone.

The ABL is of the most important with respect to the emission, transport and dispersion of airborne pollutants. The part of the ABL between the Earth's surface and the bottom of the inversion layer is known as the mixing layer. Almost all of the airborne pollutants emitted into the ambient atmosphere are transported and dispersed within the mixing layer. Some of the emissions penetrate the inversion layer and enter the free troposphere above the ABL.

In summary, the layers of the Earth's atmosphere from the surface of the ground upwards are: the ABL made up of the mixing layer capped by the inversion layer; the free troposphere; the stratosphere; the mesosphere and others. Many atmospheric dispersion models are referred to as *boundary layer models* because they mainly model air pollutant dispersion within the ABL. To avoid confusion, models referred to as *mesoscale models* have dispersion modeling capabilities that extend horizontally up to a few hundred kilometres. It does not mean that they model dispersion in the mesosphere.

Gaussian Air Pollutant Dispersion Equation

The technical literature on air pollution dispersion is quite extensive and dates back to the 1930s and earlier. One of the early air pollutant plume dispersion equations was derived by Bosanquet and Pearson. Their equation did not assume Gaussian distribution nor did it include the effect of ground reflection of the pollutant plume.

Sir Graham Sutton derived an air pollutant plume dispersion equation in 1947 which did include the assumption of Gaussian distribution for the vertical and crosswind dispersion of the plume and also included the effect of ground reflection of the plume.

Under the stimulus provided by the advent of stringent environmental control regulations, there was an immense growth in the use of air pollutant plume dispersion calculations between the late 1960s and today. A great many computer programs for calculating the dispersion of air pollutant emissions were developed during that period of time and they were called "air dispersion models". The basis for most of those models was the Complete Equation For Gaussian Dispersion Modeling Of Continuous, Buoyant Air Pollution Plumes shown below:

$$C = \frac{Q}{u} \cdot \frac{f}{\sigma_y \sqrt{2\pi}} \cdot \frac{g_1 + g_2 + g_3}{\sigma_z \sqrt{2\pi}}$$

where:

f = Crosswind dispersion parameter

$= \exp[-y^2 / (2\sigma_y^2)]$

g = Vertical dispersion parameter = $g_1 + g_2 + g_3$

g_1 = Vertical dispersion with no reflections

$= \exp[-(z - H)^2 / (2\sigma_z^2)]$

g_2 = Vertical dispersion for reflection from the ground

$= \exp[-(z + H)^2 / (2\sigma_z^2)]$

g_3 = Vertical dispersion for reflection from an inversion aloft

$$= \sum_{m=1}^{\infty} \{\exp[-(z - H - 2mL)^2 / (2\sigma_z^2)] + \exp[-(z + H + 2mL)^2 / (2\sigma_z^2)] + \exp[-(z + H - 2mL)^2 / (2\sigma_z^2)] + \exp[-(z - H + 2mL)^2 / (2\sigma_z^2)]\}$$

C = Concentration of emissions, in g/m^3, at any receptor located:

X meters downwind from the emission source point

Y meters crosswind from the emission plume centerline

Z meters above ground level

Q = Source pollutant emission rate, in g/s

u = Horizontal wind velocity along the plume centerline, m/s

H = Height of emission plume centerline above ground level, in m

σ_z = Vertical standard deviation of the emission distribution, in m

σ_y = Horizontal standard deviation of the emission distribution, in m

L = Height from ground level to bottom of the inversion aloft, in m

exp = The exponential function

The above equation not only includes upward reflection from the ground, it also includes downward reflection from the bottom of any inversion lid present in the atmosphere.

The sum of the four exponential terms in g_3 converges to a final value quite rapidly. For most cases, the summation of the series with m = 1, m = 2 and m = 3 will provide an adequate solution.

σ_z and σ_y are functions of the atmospheric stability class (i.e., a measure of the turbulence in the ambient atmosphere) and of the downwind distance to the receptor. The two most important variables affecting the degree of pollutant emission dispersion obtained are the height of the emission source point and the degree of atmospheric turbulence. The more turbulence, the better the degree of dispersion.

Equations for σ_y and σ_z are:

$$\sigma_y\ (x) = \exp(I_y + J_y \ln(x) + K_y[\ln(x)]^2)$$

$$\sigma_z\ (x) = \exp(I_z + J_z \ln(x) + K_z[\ln(x)]^2)$$

(units of σ_z, and σ_y, and x are in meters).

Coefficient	A	B	C	D	E	F
R_y	0.443	0.324	0.216	0.141	0.105	0.071
r_y	0.894	0.894	0.894	0.894	0.894	0.894
I_y	-1.104	-1.634	-2.054	-2.555	-2.754	-3.143
J_y	0.9878	1.0350	1.0231	1.0423	1.0106	1.0148
K_y	-0.0076	-0.0096	-0.0076	-0.0087	-0.0064	-0.0070
I_z	4.679	-1.999	-2.341	-3.186	-3.783	-4.490
J_z	-1.7172	0.8752	0.9477	1.1737	1.3010	1.4024
K_z	0.2770	0.0136	-0.0020	-0.0316	-0.0450	-0.0540

The classification of stability class is proposed by F. Pasquill. The six stability classes are referred to: A-extremely unstable B-moderately unstable C-slightly unstable D-neutral E-slightly stable F-moderately stable.

The resulting calculations for air pollutant concentrations are often expressed as an air pollutant concentration contour map in order to show the spatial variation in contaminant levels over a wide area under study. In this way the contour lines can overlay sensitive receptor locations and reveal the spatial relationship of air pollutants to areas of interest.

Whereas older models rely on stability classes for the determination of σ_y and σ_z, more recent models increasingly rely on the Monin-Obukhov similarity theory to derive these parameters.

Briggs Plume Rise Equations

The Gaussian air pollutant dispersion equation requires the input of *H* which is the pollutant plume's centerline height above ground level—and H is the sum of H_s (the actual physical height of the pollutant plume's emission source point) plus ΔH (the plume rise due the plume's buoyancy).

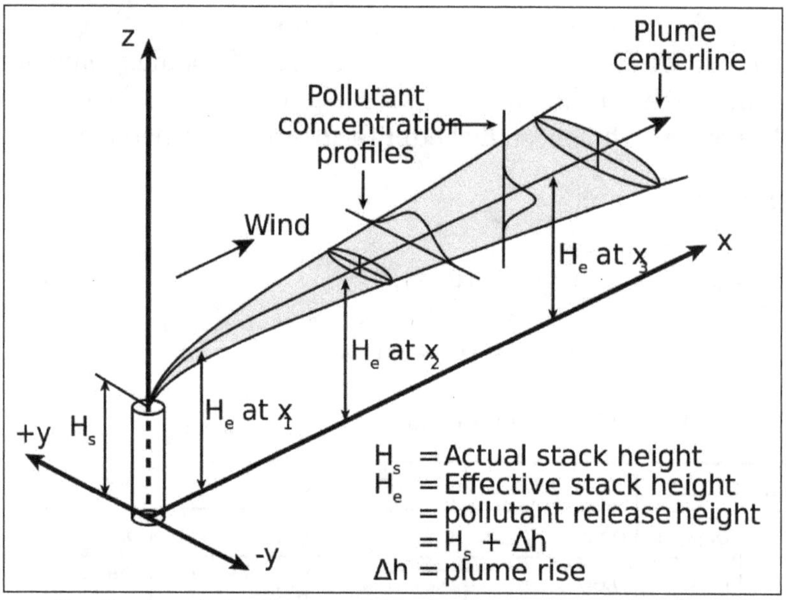

Visualization of a buoyant Gaussian air pollutant dispersion plume.

To determine ΔH, many if not most of the air dispersion models developed between the late 1960s and the early 2000s used what are known as "the Briggs equations." G.A. Briggs first published his plume rise observations and comparisons in 1965. In 1968, at a symposium sponsored by CONCAWE (a Dutch organization), he compared many of the plume rise models then available in the literature. In that same year, Briggs also wrote the section of the publication edited by Slade dealing with the comparative analyses of plume rise models. That was followed in 1969 by his classical critical review of the entire plume rise literature, in which he proposed a set of plume rise equations which have become widely known as "the Briggs equations". Subsequently, Briggs modified his 1969 plume rise equations in 1971 and in 1972.

Briggs divided air pollution plumes into these four general categories:

- Cold jet plumes in calm ambient air conditions.
- Cold jet plumes in windy ambient air conditions.
- Hot, buoyant plumes in calm ambient air conditions.
- Hot, buoyant plumes in windy ambient air conditions.

Briggs considered the trajectory of cold jet plumes to be dominated by their initial velocity momentum, and the trajectory of hot, buoyant plumes to be dominated by their buoyant momentum to the extent that their initial velocity momentum was relatively unimportant. Although Briggs proposed plume rise equations for each of the above plume categories, it is important to emphasize that "the Briggs equations" which become widely used are those that he proposed for bent-over, hot buoyant plumes.

In general, Briggs's equations for bent-over, hot buoyant plumes are based on observations and data involving plumes from typical combustion sources such as the flue gas stacks from steam-generating boilers burning fossil fuels in large power plants. Therefore, the stack exit velocities were probably in the range of 20 to 100 ft/s (6 to 30 m/s) with exit temperatures ranging from 250 to 500 °F (120 to 260 °C).

A logic diagram for using the Briggs equations to obtain the plume rise trajectory of bent-over buoyant plumes is presented below:

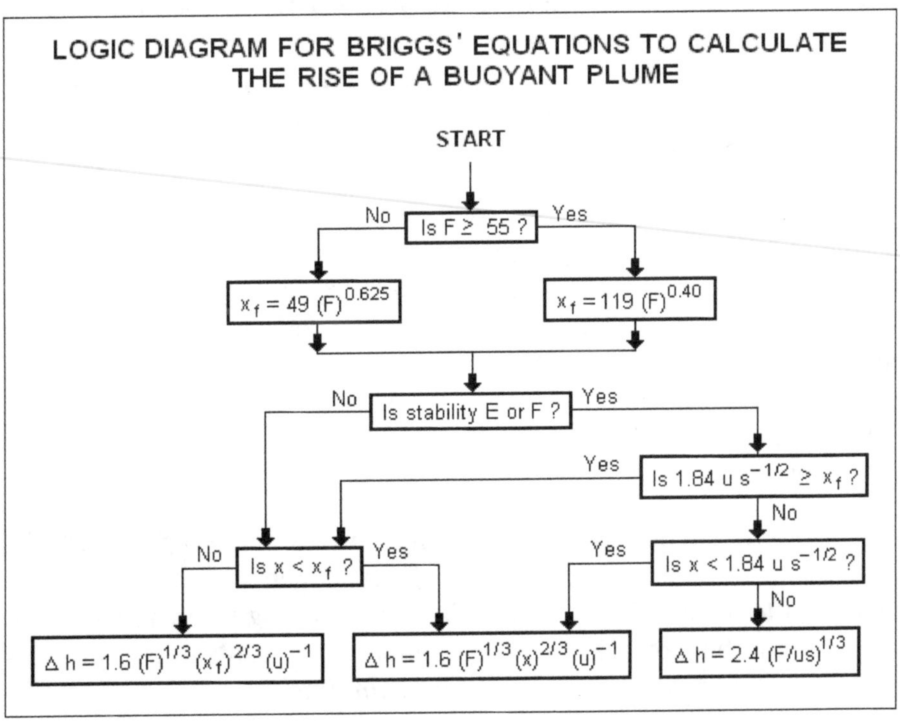

where:

Δh = Plume rise, in m,

F = Buoyancy factor, in m4s−3,

x = Downwind distance from plume source, in m,

xf = Downwind distance from plume source to point of maximum plume rise, in m,

u = Windspeed at actual stack height, in m/s,

s = Stability parameter, in s−2.

PASSIVE SAMPLING

Passive sampling is an environmental monitoring technique involving the use of a collecting medium, such as a man-made device or biological organism, to accumulate chemical pollutants in the environment over time. This is in contrast to grab sampling, which involves taking a sample directly from the media of interest at one point in time. In passive sampling, average chemical concentrations are calculated over a device's deployment time, which avoids the need to visit a sampling site multiple times to collect multiple representative samples. Currently, passive samplers have been developed and deployed to detect toxic metals, pesticides, pharmaceuticals, radionuclides, polycyclic aromatic hydrocarbons (PAHs), polychlorinated biphenyls (PCBs), and other organic compounds in water, while some passive samplers can detect hazardous substances in the air.

The underlying principle of passive sampling is the flow of contaminant molecules or ions from the sampling medium (air or water) onto a collecting medium (the passive sampler), due to Fick's first law of diffusion and, depending on the passive sampler, a greater binding affinity of contaminants with the collecting medium as compared to the sampling medium. As a result, contaminants concentrate on the collecting medium over time until they reach equilibrium with the surrounding medium. The use of passive sampling provides time-averaged concentrations of contaminants over the sampler's deployment period.

It is important to distinguish passive sampling from *active sampling*, which has the same underlying principle but employs moving parts, such as pumps, to force the sampling medium onto a collecting medium. Passive sampling solely relies on molecular diffusion and the sorbing or binding of contaminants to agents in the samplers, which is why passive sampling is also called *diffusive sampling*.

Passive sampling is also distinct from grab sampling, which is the collection of an air, water, or soil sample to analyze directly for contaminants. These samples represent a single point in time and provide information about contaminant concentration at one point in time, unlike passive sampling devices or organisms.

Many different kinds of passive samplers have been developed and have sampled many different contaminants, including:

- Antibiotics.
- Bisphenols.
- Dioxins.
- Flame retardants.
- Furans.
- Nanoparticles.
- Oxyanions such as phosphates.
- Polybrominated diphenyl ethers (PBDEs).
- Polycyclic aromatic hydrocarbons (PAHs).
- Polychlorinated biphenyls (PCBs).
- Pesticides.
- Pharmaceuticals.
- Radionuclides.
- Toxic metals (Such as copper, zinc, and lead).

Passive Sampling in Water

Several kinds of passive sampling devices exist for monitoring pollutants present in water. In addition to these devices, organisms, such as mussels, living in the environment also "passively sample" contaminants (bioaccumulation) and can be used to monitor water pollution (biomonitoring).

Chemcatcher

Chemcatcher can passively sample inorganic pollutants (metals) and a wide range of organic pollutants. It is composed of a single-use disc, with or without a membrane, sealed into a plastic support. Types of receiving phases and membranes vary greatly, depending on the target chemicals to be sampled. Time-averaged water concentrations of many chemical pollutants can be determined as long as sampling rate and the water flow rate are known.

A Chemcatcher deployment kit with three sampling discs.

Diffusive Gradients in Thin Films (DGT)

Diffusive gradients in thin films (DGT) samplers passively sample ionic trace metals, as well as antibiotics, oxyanions, bisphenols, and nanoparticles in different configurations. They are composed of plastic pistons and caps, with a window that exposes a binding gel, diffusive gel, and filter membrane to the sampling water. They can be used in both freshwater and marine environments, as well as in the water located between freshwater and marine sediment particles, called *pore water* or *interstitial water*. Once the mass of accumulated contaminants on the DGT sampler is known, the DGT equation (based on Fick's law) can be used to calculate the time averaged water concentration of contaminants.

Peepers

Peepers are passive diffusion samplers used for metals in freshwater and marine sediment pore water, so they can be used to find areas that may have metal-contaminated sediments. Peepers are plastic vessels filled with clean water and covered in a dialysis membrane, which allows metals in sediment pore water to enter the water inside the peeper. They are usually placed deep enough into sediment to be in an anoxic environment, in which metals will be soluble enough to sample. If the peepers are deployed long enough so the sediment pore water and contained peeper water reach equilibrium, they can accurately provide metal concentrations in sampled sediment pore water.

Polar Organic Chemical Integrative Sampler (POCIS)

Polar organic chemical integrative samplers (POCIS) sample polar organic contaminants with a log octanol-water partition coefficient (K_{ow}) value that is less than 3.

Examples of these types of chemicals include polar pesticides, pharmaceuticals, illicit drugs, flame retardants, and drug metabolites. The POCIS is composed of variable numbers of solid sorbent discs attached to a support rod and encased in a metal cage, and has two possible sorbent configurations, the pesticide-POCIS and pharmaceutical-POCIS. As long as the amount of water passing over the sampler is known, polar organic contaminant water concentrations can be calculated after extracting sorbed contaminants from a POCIS.

Semipermeable Membrane Devices (SPMDs)

Semipermeable membrane devices (SPMDs) passively sample nonpolar organic contaminants with a log octanol-water partition coefficient (K_{ow}) value greater than 3. Examples of these types of chemicals include polycyclic aromatic hydrocarbons (PAHs), polychlorinated biphenyls (PCBs), polybrominated diphenyl ethers (PBDEs), chlorinated pesticides, dioxins, and furans. SPMDs are composed of sealed plastic tubing filled with triolein, in which nonpolar organics are very soluble and which serves as a representation of the fatty tissues of aquatic organisms. The tubing is then weaved between metal rods and enclosed in a metal cage. The sampler can be made in varying lengths of tubing for different applications, since sampling rate depends on the surface area of tubing exposed to the water. As long as the amount of water passing over the sampler is known, nonpolar organic contaminant water concentrations can be calculated after extracting contaminants from a SPMD.

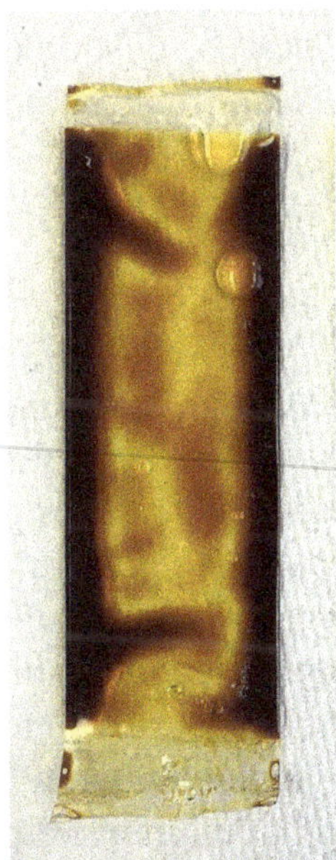

A 7.5 centimeter SLMD, filled with a 1:1 mixture of Kelex-100 and oleic acid.

Stabilized Liquid Membrane Devices (SLMDs)

Stabilized liquid membrane devices (SLMDs) passively sample ionic metals in freshwater. They are made of low-density polyethylene plastic tubing sections that are sealed on both ends and filled with an equal mixture of oleic acid and metal chelating agent. They work by interacting with calcium and magnesium ions in freshwater, which forms a hydrophobic film on the outside the SLMD plastic membrane in which the chelating agent can bind to metals in the sampling water. They have been deployed for up to month-long periods in the field, alone or covered by a plastic tube housing to mediate water flow. Metal weight accumulated by a SLMD over its deployment period can be calculated and divided by the SLMD deployment time to get an average metal weight accumulated per time unit, but currently, no method has been developed to convert this to an average metal concentration. In addition, SLMD sampling rates greatly vary with water flow rate, which plastic housings can be used to control.

Passive Sampling in Air

Passive sampling can also be accomplished for contaminants in the air, including airborne particles and hazardous vapors and gases. This can be done with man-made devices or with biomonitoring organisms, such as lichens.

Sorbent Tubes

Sorbent tubes are passive samplers for volatile organic compounds (VOCs). They are glass tubes filled with adsorbing materials, such as charcoal or silica gel, which the air to be sampled passes through. The adsorbing materials remove VOCs from the air that passes through them, and the VOCs can be desorbed and analyzed. Air concentrations can be calculated using the amount of air that flowed through the sorbent tube and the amount of contaminants desorbed.

Advantages

Contaminant concentrations from passive sampling reflect average contamination throughout the sampler deployment time, meaning the sample will capture contaminant concentration fluctuations over the whole deployment period. Traditional grab sampling does not do this, since collected samples only represent a single moment in time and multiple grab samples must be taken to observe variation in contaminant concentrations over time. This integrative sampling method can also can result in the detection of chemicals present at such low concentrations that they would be undetected in a grab sample, due to concentration of the chemicals on the sampler over time. As a result, passive sampling has the potential to be a less time-intensive, less expensive and more accurate sampling method than grab sampling.

In addition, passive samplers are often easy to use and deploy, have no pumps or moving parts, and do not require electricity, since they rely on the molecular diffusion of contaminants or binding of contaminants to agents within the samplers, unlike active sampling. They may also be inexpensive and simple to construct, such as SLMDs, which only require sealed plastic tubing and two chemical components.

Passive sampling may also more accurately reflect metal concentrations that are bioavailable to organisms than other sampling methods. For example, the SPMD sampler uses a semipermeable membrane and triolein (a triglyceride), both of which mimic toxicant uptake by organism fatty tissue. However, this depends on the type of passive sampler used, since some samplers, such as peepers, rely solely on molecular diffusion, which may be too simple to be reflective of the complex processes of contaminant uptake in an organism.

Disadvantages

Since passive sampling provides information about average contaminant concentrations, all possible concentrations over the sampler deployment time are included in this average value. However, there is no way of finding out the complete range of contaminant concentrations over the deployment time at a single site with only passive sampling. If high and low concentrations of contaminants throughout the sampling period are needed, other sampling methods should be used in conjunction with passive sampling.

Not all passive samplers have universally accurate ways to convert contaminant masses accumulated into water concentrations, which are used in government regulation, such as the United States Environmental Protection Agency Clean Water Act. With some samplers, as with the DGT, this can be done using equations developed for the samplers, but not all samplers have these. Passive sampler deployment time is also limited depending on the sampler's capacity; for example, SLMDs have been deployed for month-long periods, but may reach saturation and stop sampling much sooner if metal concentrations and water flow rates are high enough. However, this issue is avoidable if literature on the relevant passive sampler is examined for background information about sampler capacity and ideal deployment times prior to deployment.

PARTICIPATORY MONITORING

Scanning the sea off Greenland for seabirds as part of Greenland's documentation and management system PISUNA, a participatory monitoring programme.

Participatory monitoring (also known as collaborative monitoring, community-based monitoring, locally based monitoring, or volunteer monitoring) is the regular collection of measurements or other kinds of data (monitoring), usually of natural resources and biodiversity, undertaken by local residents of the monitored area, who rely on local natural resources and thus have more local knowledge of those resources. Those involved usually live in communities with considerable social cohesion, where they regularly cooperate on shared projects.

Participatory monitoring has emerged as an alternative or addition to professional scientist-executed monitoring. Scientist-executed monitoring is often costly and hard to sustain, especially in those regions of the world where financial resources are limited. Moreover, scientist-executed monitoring can be logistically and technically difficult and is often perceived to be irrelevant by resource managers and the local communities. Involving local people and their communities in monitoring is often part of the process of sharing the management of land and resources with the local communities. It is connected to the devolution of rights and power to the locals. Aside from potentially providing high-quality information, participatory monitoring can raise local awareness and build the community and local government expertise that is needed for addressing the management of natural resources.

Participatory monitoring is sometimes included in terms such as citizen science, crowd-sourcing, 'public participation in scientific research' and participatory action research.

The term 'participatory monitoring' embraces a broad range of approaches, from self-monitoring of harvests by local resource users themselves, to censuses by local rangers, and inventories by amateur naturalists. The term includes techniques labelled as 'self-monitoring', ranger-based monitoring', 'event-monitoring', 'participatory assessment, monitoring and evaluation of biodiversity', 'community-based observing', and 'community-based monitoring and information systems'.

Many of these approaches are directly linked to resource management, but the entities being monitored vary widely, from individual animals and plants, through habitats, to ecosystem goods and services. However, all of the approaches have in common that the monitoring is carried out by individuals who live in the monitored places and rely on local natural resources, and that local people or local government staff are directly involved in formulation of research questions, data collection, and (in most instances) data analysis, and implementation of management solutions based on research findings.

Participatory monitoring is included in the term 'participatory monitoring and management' which has been defined as "approaches used by local and Indigenous communities, informed by traditional and local knowledge, and, increasingly, by contemporary science, to assess the status of resources and threats on their land and advance sustainable economic opportunities based on the use of natural resources". term 'participatory monitoring and management' is particularly used in tropical, Arctic and developing regions, where communities are most often the custodians of valuable biodiversity and extensive natural ecosystems.

Other definitions for participatory monitoring have also been proposed, including:

- "The systematic collection of information at regular intervals for initial assessment and for the monitoring of change. This collection is undertaken by locals in a community who do not have professional training".

Likewise, the term community-based monitoring of natural resources' has been defined as:

- "A process where concerned citizens, government agencies, industry, academia, community groups and local institutions collaborate to monitor, track, and respond to issues of common community concern".
- "Monitoring of natural resources undertaken by local stakeholders using their own resources and in relation to aims and objectives that make sense to them".
- "A process of routinely observing environmental or social phenomena, or both, that is led and undertaken by community members and can involve external collaboration and support of visiting researchers and government agencies".

Limitations

It has been suggested that participatory monitoring is unlikely to provide quantitative data on large-scale changes in habitat area, or on populations of cryptic species that are hard to identify or census reliably. It has also been suggested that participatory monitoring is not suitable for monitoring resources that are so valuable they attract powerful outsiders. Likewise, in areas where changes, threats, or interventions operate in complex fashions, where rural people do not depend on the use of natural resources and there are no real benefits flowing to the local people from doing monitoring work (or the costs to local people of involvement exceed the benefits), or where there is a poor relationship between the authorities and the local people, participatory monitoring is probably less likely to yield useful data and management solutions than conventional scientific approaches.

Approaches

Thematically, participatory monitoring has considerable potential in several areas, including:

- For connecting knowledge systems: In efforts to bring Indigenous and local knowledge systems into the science–policy interface such as the Intergovernmental Platform for Biodiversity and Ecosystem Services.

- For monitoring rapidly changing environments: To inform resource management in rapidly changing environments such as the Arctic, where Indigenous and local communities have detailed knowledge of key components of their environment, such as sea-ice, snow, weather patterns, caribou and other natural resources.
- In Payment for Ecosystem Services (PES) programs: To connect environmental performance with payment schemes such as REDD+.
- For reinforcing international agreements: In efforts to link international environmental agreements to decision-making in the 'real world'.

Typology

A typology of monitoring schemes has been proposed, determined on the basis of relative contributions of local stakeholders and professional researchers, and supported by findings from statistical analysis of published schemes. The typology identified 5 categories of monitoring schemes that between them span the full spectrum of natural resource monitoring protocols:

Category A. Autonomous Local Monitoring: In this category the whole monitoring process—from design, to data collection, to analysis, and finally to use of data for management decisions—is carried out autonomously by local stakeholders. There is no direct involvement of external agencies.

Category B. Collaborative Monitoring with Local Data Interpretation: In these schemes, the original initiative was taken by scientists but local stakeholders collect, process and interpret the data, although external scientists may provide advice and training. The original data collected by local people remain in the area being monitored, which helps create local ownership of the scheme and its results, but copies of the data may be sent to professional researchers for in-depth or larger-scale analysis.

Category C. Collaborative Monitoring with External Data Interpretation: The third most distinct group is monitoring scheme category C. These schemes were designed by scientists who also analyse the data, but the local stakeholders collect the data, take decisions on the basis of the findings and carry out the management interventions emanating from the monitoring scheme.

Category D. Externally Driven Monitoring with Local Data Collectors: This category of monitoring scheme involves local stakeholders only in data collection. The design, analysis, and interpretation of the monitoring results are undertaken by professional researchers—generally far from the site. Monitoring schemes of category D are mostly long-running 'citizen science' projects from Europe and North America.

Category E. Externally Driven, Professionally Executed Monitoring: Monitoring schemes of category E do not involve local stakeholders. Design of the scheme, analysis of the results, and management decisions derived from these analyses are all undertaken by professional scientists funded by external agencies.

The use of Technology for Participatory Monitoring

Traditional methods of data collection for participatory monitoring use paper and pen. This has advantages in terms of low cost of materials and training, simplicity, and reduced potential for technical hitches. However, all data must be transcribed for analysis, which takes time and can be subject to transcription errors. Increasingly, participatory monitoring initiatives incorporate technology, from GPS recorders to georeference the data collected on paper, to drones to survey remote areas, phones to send simple reports via SMS, or smartphones to collect and store data. Various apps exist to create and manage data collection forms on smartphones (e.g. Open Data Kit, Sapelli and others).

Some initiatives find that the use of smartphones for data collection has advantages over paper-based systems. The advantages include that very little equipment need be carried on a survey, a large amount and variety of data can be stored (geographical locations, photos and audio, as well as data entered onto monitoring forms) and data can be shared rapidly for analysis without transcription errors. The use of smartphones can incentivise young people to get involved in monitoring, sparking an interest in conservation. Some apps are especially designed to be usable by illiterate monitors. If local people risk threats or violence by monitoring illegal activities, the true purpose of the phones can be denied, and the monitoring data locked away. However, phones are expensive; are vulnerable to damage and technical issues; necessitate additional training - not least due to rapid technological change; phone charging can be a challenge (especially under thick forest canopies); and uploading data for analysis is difficult in areas without network connections.

Data Sharing in Participatory Monitoring

A key challenge for participatory monitoring is to develop ways to store, manage and share data and to do this in ways that respect the rights of the communities that supplied the data. A 'rights-based approach to data sharing' can be based on principles of free, prior and informed consent, and prioritise the protection of the rights of those who generated the data, and/or those potentially affected by data-sharing. Local people can do much more than simply collect data: they can also define the ways that this data is used, and who has access to it.

Clear agreements on data sharing are especially important for initiatives where diverse data is collected, of variable relevance to different stakeholders. For example, monitoring could on the one hand, investigate sensitive social problems within a community, or contested resources at the centre of local conflicts or illegal exploitation - data that community leaders might want to keep confidential and address locally; on the other hand, the same initiative could generate data on forest biomass, of greater interest to external stakeholders.

One way to establish the rules around data sharing is to set up a data sharing protocol. This can define:

- The infrastructure for data storage and management (computer programmes, hard drives and cloud storage). Local capacity should be strong enough to access, manage and retain control of the data.
- Data classification: Discussions in the communities can set out how different types of data can be used – for example a traffic light system can define 'red' data that is confidential to the community, 'amber' data which should be discussed prior to any use, and 'green' data that is approved for release.
- Processes for data sharing: This defines the roles and responsibilities of different people, and the processes to be followed for requests to access data, dependent on how that data is classified.
- Reporting: The protocol can set out how data should be reported, for example specifying the manner and frequency with which findings are reported to the local community, and ensuring that technical data is presented in a way that is compatible with external systems (e.g. government databanks or processes to respond to findings).

BUFFER STRIP

Contour buffer strips used to retain soil and reduce erosion.

A buffer strip is an area of land maintained in permanent vegetation that helps to control air, soil, and water quality, along with other environmental problems, dealing primarily on land that is used in agriculture. Buffer strips trap sediment, and enhance filtration of nutrients and pesticides by slowing down runoff that could enter the local surface waters. The root systems of the planted vegetation in these buffers hold soil particles together which alleviate the soil of wind erosion and stabilize stream banks providing protection against substantial erosion and landslides. Farmers can also use buffer strips to square up existing crop fields to provide safety for equipment while also farming more efficiently.

Buffer strips can have several different configurations of vegetation found on them varying from simply grass to combinations of grass, trees, and shrubs. Areas with diverse vegetation provide more protection from nutrient and pesticide flow and at the same time provide better biodiversity amongst plants and animals.

Many country, state, and local governments provide financial incentives for conservation programs such as buffer strips because they help stabilize the environment even when the land is being used. Buffer strips not only stabilize the land but can also provide a visual demonstration that land is under stewardship.

Types

Buffers within Fields

A grassed waterway reduces soil erosion and captures most nutrients and pesticides that would normally wash out of crop fields and into major waters. These waterways help to carry surface water at a non-erosive velocity to an area where it will have a stable outlet. Outlets must be adequate enough to allow water to drain without ponding or flooding the area being protected, while also preventing erosion of the water into the outlet which can be accomplished through the use of riprap. A limitation is during large runoff events, when soil is saturated, grassed waterways will have a very concentrated flow of water making them not as effective during high rainfalls. Grassed waterways require very little maintenance once they are introduced with major upkeep being mowing of the grass and reseeding. Farm machinery and cattle can cross these waterways but it may be hazardous during wet periods. One of the major disadvantages of waterways are actually getting them established. A late summer or early fall seeding when rainfall is minimal is recommended to allow the seed to have the best chance at establishing a root system.

Contour buffer strips are strips of perennial vegetation alternated with wider cultivated strips of cropland. This type of buffer strip is most effective when runoff water enters uniformly as sheetflow. They are very adapted to trapping pesticides and reducing rill erosion. These buffers need to be at least 15 feet (4.6 m) wide and make up for 20 to 30 percent of slope of an area. A lot of the time contour buffers can be used as a very inexpensive substitute for terraces. Most of the time a grass is selected that can be harvested during mid-summer. These buffers are not permanent and can be moved up and down hillsides from season to season in order to re-establish vegetation.

Edge of Field Buffers

A riparian buffer of vegetation lining a farm creek.

Vegetative barriers are narrower buffer strips of hardy, native, perennial grasses or shrubs planted in parallel rows to crops. They are very effective in reducing wind and water erosion which results in sediment trapping and water infiltration. They function in similar fashion to a contour buffer strip, just much narrower.

Field borders are bands or strips of perennial vegetation that is found on the edge of a cropland field. Field borders help with runoff only when it flows over the strip. They're very effective in benefiting spraying operations because they allow for extra room between adjacent fields. They also provide room for farming equipment to turn around. Field borders are effective in reducing wind and water erosion and provide great wildlife habitat.

Filter strips are areas of grass or other permanent vegetation that protect riparian areas from sediment runoff, pesticides, pathogens, organics and nutrients. These strips are very important in protecting stream banks and water quality. Filter strips work best when other conservation practices are used in order to drain water in their direction. Filter strips were originally used mostly in agriculture, but now are a common practice in urban areas where water quality has become an increasingly important issue.

Riparian forest buffers are diverse communities of trees, shrubs and native perennial grasses. They are great for providing habitat for wildlife on land and in the water. Taller trees next to the streams help to lower water temperatures with shade which improves aquatic communities. The shrubs and grasses help to slow flooding and the larger trees can sometimes intercept nitrates before they reach the water with their deep roots.

Wind Buffers

The main purpose of a windbreak or a shelterbelt is to protect areas from wind causing erosion on the bare soil of croplands. Windbreaks can also serve as an area that separates fields and protects them from spray drift of pesticides.

Wildlife Habitat

Buffer strips are very important in helping to provide habitat for many species of wildlife in the open farm lands by causing an edge effect. With much of the land open on farms having an edge allows a safe-haven for animals to move between different ecosystems. Buffers are also helpful in conserving biodiversity especially to that of rare or endangered species through the incorporation of native grasses into their seeding by the landowner. Native grasses are especially important to pheasants, quail, chukar and songbirds because they provide the foods they need as well as the shelter from predators. Since most buffer strip areas have limited disturbance from farming it allows for a shelter to hide year round for many of the species including insects, birds, and mammals. When buffers run into and follow riparian areas along stream beds it is important to have larger vegetation like trees and shrubs that shade the water from the open sun. The water is then able to be cooler allowing for greater fish production and other aquatic plants and other biota to thrive in a less disturbed environment.

This area of vegetation following a body of water is noted as a riparian zone. These buffer areas often incorporate large trees that protect stream banks from excessive erosion and shade aquatic areas. The shade provided by the larger trees reduces water temperatures and light intensity from ultraviolet light. Debris including leaves and branches that fell from trees, often contain aquatic invertebrates important to the structure of the water following the "River Continuum" concept. Since the riparian zones contain a larger variety of plants the overall diversity is much greater as well. With more photosynthesis and higher amounts of available water plant primary production can increase in turn creating more potential food for the wildlife.

Planted Vegetation

The USDA shows that grassed waterways of less than 5% slopes for a chosen waterway will greatly reduce velocity of draining water within the land. The Natural Resources Conservation Service (NRCS) offers the use of an online Soil survey to view the area of land to be planted and examine slope and soil drainage. As viewed by the Natural Resources Conservation Service, soil drainage is the number one priority for location of certain grasses. Poorer drainage causes less infiltration of water into the soil as groundwater recharge causing ponding and flooding of surrounding crops. Higher drainage becomes more droughty which hinders the establishment of certain grasses good for waterways. Medium soil drainage becomes the most suitable for planting. Determining exact grasses and seed amounts to plant follows specifications of a local NRCS Field Office Technical Guide and the Contour Buffer practice standard while taking practice on land and regional environmental conditions into consideration.

The NRCS has shown contour strip cropping provides the greatest effectiveness when slopes of the area are within 4-8%. For greater success in lowering the erosion, contour strips on the slopes need to follow the contours of the landscape. Row crops like corn, legumes, or soybeans should alternate strips of small grains or forages to successfully limit erosion and slowing or capturing the runoff of fertilizers and pesticides. Tests from the NRCS have shown that the smaller tighter rooted grasses and hays provide more surface cover to prevent rain and wind erosion while slowing runoff, increasing infiltration, and trapping sediment from the high erosive row crops. Proper planting of the contour crops is important for higher success of erosion control to protect highly fertile topsoil.

SCRUBBER

Scrubber systems (e.g. chemical scrubbers, gas scrubbers) are a diverse group of air pollution control devices that can be used to remove some particulates and/or gases from industrial exhaust streams. The first air scrubber was designed to remove carbon dioxide from the air of an early submarine, the Ictineo I, a role for which they continue to be used today. Traditionally, the term "scrubber" has referred to pollution control devices that use liquid to wash unwanted pollutants from a gas stream. Recently, the term has also been used to describe systems that inject a dry reagent or slurry into a dirty exhaust stream to "wash out" acid gases. Scrubbers are one of the primary devices that control gaseous emissions, especially acid gases. Scrubbers can also be used for heat recovery from hot gases by flue-gas condensation. They are also used for the high flows in solar, PV, or LED processes.

There are several methods to remove toxic or corrosive compounds from exhaust gas and neutralize it.

Combustion

Combustion is sometimes the cause of harmful exhausts, but, in many cases, combustion may also be used for exhaust gas cleaning if the temperature is high enough and enough oxygen is available.

Wet Scrubbing

The exhaust gases of combustion may contain substances considered harmful to the environment, and the scrubber may remove or neutralize those. A wet scrubber is used for cleaning air, fuel gas or other gases of various pollutants and dust particles. Wet scrubbing works via the contact of target compounds or particulate matter with the scrubbing solution. Solutions may simply be water (for dust) or solutions of reagents that specifically target certain compounds.

Process exhaust gas can also contain water-soluble toxic and/or corrosive gases like hydrochloric acid (HCl) or ammonia (NH_3). These can be removed very well by a wet scrubber.

Removal efficiency of pollutants is improved by increasing residence time in the scrubber or by the increase of surface area of the scrubber solution by the use of a spray nozzle, packed towers or an aspirator. Wet scrubbers may increase the proportion of water in the gas, resulting in a visible stack plume, if the gas is sent to a stack.

Wet scrubbers can also be used for heat recovery from hot gases by flue-gas condensation. In this mode, termed a condensing scrubber, water from the scrubber drain is circulated through a cooler to the nozzles at the top of the scrubber. The hot gas enters the scrubber at the bottom. If the gas temperature is above the water dew point, it is initially cooled by evaporation of water drops. Further cooling cause water vapors to condense, adding to the amount of circulating water.

The condensation of water releases significant amounts of low temperature heat due to the high value of the specific latent heat of the vaporisation of water (more than 2 gigajoules (560 kWh) per ton of water), that can be recovered by the cooler for e.g. district heating purposes.

Excess condensed water must continuously be removed from the circulating water. The gas leaves the scrubber at its dew point, so even though significant amounts of water may have been removed from the cooled gas, it is likely to leave a visible stack plume of water vapor.

Dry Scrubbing

A dry or semi-dry scrubbing system, unlike the wet scrubber, does not saturate the flue gas stream that is being treated with moisture. In some cases no moisture is added, while in others only the amount of moisture that can be evaporated in the flue gas without condensing is added. Therefore, dry scrubbers generally do not have a stack steam plume or wastewater handling/disposal requirements. Dry scrubbing systems are used to remove acid gases (such as SO_2 and HCl) primarily from combustion sources.

There are a number of dry type scrubbing system designs. However, all consist of two main sections or devices: A device to introduce the acid gas sorbent material into the gas stream and a particulate matter control device to remove reaction products, excess sorbent material as well as any particulate matter already in the flue gas.

Dry scrubbing systems can be categorized as dry sorbent injectors (DSIs) or as spray dryer absorbers (SDAs). Spray dryer absorbers are also called semi-dry scrubbers or spray dryers.

Dry scrubbing systems are often used for the removal of odorous and corrosive gases from wastewater treatment plant operations. The medium used is typically an activated alumina compound impregnated with materials to handle specific gases such as hydrogen sulfide. Media used can be mixed together to offer a wide range of removal for other odorous compounds such as methyl mercaptans, aldehydes, volatile organic compounds, dimethyl sulfide, and dimethyl disulfide.

Dry sorbent injection involves the addition of an alkaline material (usually hydrated lime, soda ash, or sodium bicarbonate) into the gas stream to react with the acid gases. The sorbent can be injected directly into several different locations: the combustion process, the flue gas duct (ahead of the particulate control device), or an open reaction chamber (if one exists). The acid gases react with the alkaline sorbents to form solid salts which are removed in the particulate control device. These simple systems can achieve only limited acid gas (SO_2 and HCl) removal efficiencies. Higher collection efficiencies can be achieved by increasing the flue gas humidity (i.e., cooling using water spray). These devices have been used on medical waste incinerators and a few municipal waste combustors.

In spray dryer absorbers, the flue gases are introduced into an absorbing tower (dryer) where the gases are contacted with a finely atomized alkaline slurry. Acid gases are absorbed by the slurry mixture and react to form solid salts which are removed by the particulate control device. The heat of the flue gas is used to evaporate all the water droplets, leaving a non-saturated flue gas to exit the absorber tower. Spray dryers are capable of achieving high (80+%) acid gas removal efficiencies. These devices have been used on industrial and utility boilers and municipal waste incinerators.

Adsorber

Many chemicals can be removed from exhaust gas also by using adsorber material. The flue gas is passed through a cartridge which is filled with one or several adsorber materials and has been adapted to the chemical properties of the components to be removed. This type of scrubber is sometimes also called dry scrubber. The adsorber material has to be replaced after its surface is saturated. adsorption is a surface phenomena, absorption involves the entire material. Ex: Activated carbon an adsorbent, used for the adsorption of odorous compounds.

Mercury Removal

Mercury is a highly toxic element commonly found in coal and municipal waste. Wet scrubbers are only effective for removal of soluble mercury species, such as oxidized mercury, Hg^{2+}. Mercury vapor in its elemental form, Hg^{o}, is insoluble in the scrubber slurry and not removed. Therefore, an additional process of Hg^{o} conversion is required to complete mercury capture. Usually halogens are added to the flue gas for this purpose. The type of coal burned as well as the presence of a selective catalytic reduction unit both affect the ratio of elemental to oxidized mercury in the flue gas and thus the degree to which the mercury is removed.

In July 2015, one study found that some mercury scrubbers installed on coal power plants inadvertently capture PAH (polycyclic aromatic hydrocarbons) emissions as well.

Scrubber Waste Products

One side effect of scrubbing is that the process only moves the unwanted substance from the exhaust gases into a liquid solution, solid paste or powder form. This must be disposed of safely, if it can not be reused.

For example, mercury removal results in a waste product that either needs further processing to extract the raw mercury, or must be buried in a special hazardous wastes landfill that prevents the mercury from seeping out into the environment.

As an example of reuse, limestone-based scrubbers in coal-fired power plants can produce a synthetic gypsum of sufficient quality that can be used to manufacture drywall and other industrial products.

Bacteria Spread

Poorly maintained scrubbers have the potential to spread disease-causing bacteria. The problem is a result of inadequate cleaning. For example, the cause of a 2005 outbreak of Legionnaires' disease in Norway was just a few infected scrubbers. The outbreak caused 10 deaths and more than 50 cases of infection.

Scrubbers on Ships

The International Maritime Organization (IMO)has adopted guidelines on the approval, installation and use of scrubbers (exhaust gas cleaning systems) onboard ships to ensure compliance with the sulphur regulation of MARPOL Annex VI. Flag States must approve such systems and port States can (as part of their port state control) ensure that such systems are functioning correctly. If a scrubber system is not functioning probably (and the IMO procedures for such malfunctions are not adhered to), port States can sanction the ship.

RAIN GAUGE

A rain gauge is a meteorological instrument to measure the precipitating rain in a given amount of time per unit area. The instrument consists of a collection container which is placed in an open area. The precipitation is measured in terms of the height of the precipitated water accumulated in the container per given time and is expressed in millimetres. Since the same amount of rain precipitation is assumed to be occurring around the container, the area of collection is not a factor. However, it should not be too small, neither should it be too large. Due to spatial uniformity of rainfall, 1 mm of measured precipitation is the equivalent of 1 L of precipitated rain water volume per metre squared.

Types of Rain Gauge

A self-recording rain gauge (interior).

Types of rain gauges include graduated cylinders, weighing gauges, tipping bucket gauges, and simply buried pit collectors. Each type has its advantages and disadvantages while collecting rain data.

Standard Rain Gauge

The standard United States National Weather Service rain gauge, developed at the start of the 20th century, consists of a funnel emptying into a graduated cylinder, 20 mm in radius, which fits inside a larger container that is 200 mm in diameter and 500 mm tall. If the rainwater overflows the graduated inner cylinder, the larger outer container will catch it. When measurements are taken, the height of the water in the small graduated cylinder is measured, and the excess overflow in the large container is carefully poured into another graduated cylinder and measured to give the total rainfall. A cone meter is sometimes used to prevent leakage that can result in alteration of the data. In locations using the metric system, the cylinder is usually marked in mm and will measure up to 250 millimetres (9.8 in) of rainfall. Each horizontal line on the cylinder is 0.5 millimetres (0.02 in). In areas still using Imperial units, each horizontal line represents 0.01 inch.

Pluviometer of Intensities

The pluviometer of intensities (or Jardi's pluviometer) is a tool that measures the average intensity of rainfall in a certain interval of time. It was initially designed to record the rainfall regime in Catalonia, but eventually spread throughout the world.

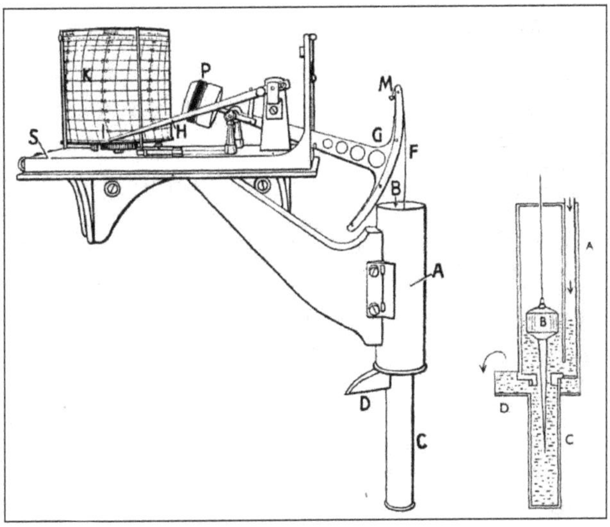

Pluviometer of intensities.

It employs the principle of feedback the incoming water pushes the buoy upwards, making the lower "adjusting conic needle" to let pass the same amount of water that enters into the container, this way the needle records on the drum the amount of water flowing through it at every moment—in mm of rainfall per square meter.

It consists of a rotating drum that rotates at constant speed, this drum drags a graduated sheet of cardboard, which has the time at the abscissa while the y-axis indicates the height of rainfall in mm of rain. This height is recorded with a pen that moves vertically, driven by a buoy, marking on the paper the rainfall over time. Each cardboard sheet is usually used for one day.

While the rain falls, the water collected by the funnel falls into the container and raises the buoy that makes the pen arm raising in the vertical axis marking the cardboard accordingly. If the rainfall does not vary, the water level in the container remains constant, and while the drum rotates, the pen's mark it is more or less a horizontal line, proportional to the amount of water that has fallen. When the pen reaches the top edge of the recording paper, it means that the buoy is "up high in the tank" leaving the tip of the conical needle in a way that uncovers the regulating hole, *i.e.*, the maximum flow that the apparatus is able to record. If the rain suddenly decreases, making the container (as it empties) to quickly lower the buoy, that movement corresponds to a steep slope line that can reach the bottom of the recorded cardboard, if it stops raining.

The rain gauge of intensities allowed precipitation to be recorded over many years, particularly in Barcelona (95 years), apart from many other places around the world, such as Hong Kong.

Weighing Precipitation Gauge

A weighing-type precipitation gauge consists of a storage bin, which is weighed to record the mass. Certain models measure the mass using a pen on a rotating drum, or by using a vibrating wire attached to a data logger. The advantages of this type of gauge over tipping buckets are that it does not underestimate intense rain, and it can measure other forms of precipitation, including rain, hail and snow. These gauges are, however, more expensive and require more maintenance than tipping bucket gauges.

The weighing-type recording gauge may also contain a device to measure the number of chemicals contained in the location's atmosphere. This is extremely helpful for scientists studying the effects of greenhouse gases released into the atmosphere and their effects on the levels of the acid rain. Some Automated Surface Observing System (ASOS) units use an automated weighing gauge called the AWPAG (All Weather Precipitation Accumulation Gauge).

Tipping Bucket Rain Gauge

The tipping bucket rain gauge consists of a funnel that collects and channels the precipitation into a small seesaw-like container. After a pre-set amount of precipitation falls, the lever tips, dumping the collected water and sending an electrical signal. An old-style recording device may consist of a pen mounted on an arm attached to a geared wheel that moves once with each signal sent from the collector. In this design, as the wheel turns the pen arm moves either up or down leaving a trace on the graph and at the same time making a loud "click".

The interior of a tipping bucket rain gauge.

The exterior of a tipping bucket rain gauge.

The tipping bucket rain gauge is not as accurate as the standard rain gauge, because the rainfall may stop before the lever has tipped. When the next period of rain begins it may take no more than one or two drops to tip the lever. This would then indicate that preset amount has fallen when only a fraction of that amount has actually fallen. Tipping buckets also tend to underestimate the amount of rainfall, particularly in snowfall and heavy rainfall events. The advantage of the tipping bucket rain gauge is that the character of the rain (light, medium, or heavy) may be easily obtained. Rainfall character is decided by the total amount of rain that has fallen in a set period (usually 1 hour) and by counting the number of pulses clicks in a 10-minute period the observer can decide the character of the rain. Algorithms may be applied to the data as a method of correcting the data for high-intensity rainfall.

Tipping bucket rain gauge recorder.

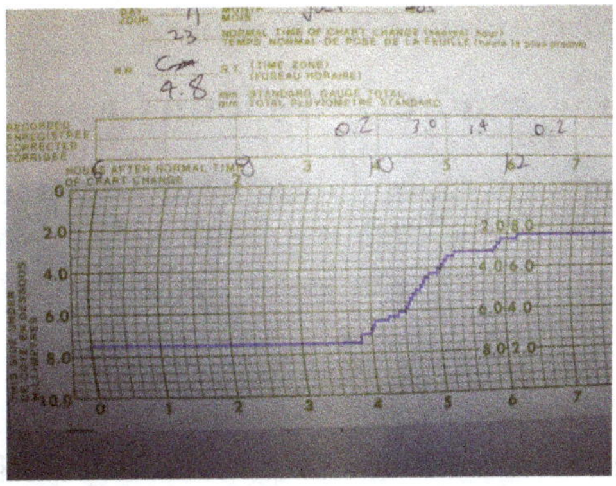

Close up of a tipping bucket rain gauge recorder chart.

Modern tipping rain gauges consist of a plastic collector balanced over a pivot. When it tips, it actuates a switch (such as a reed switch) which is then electronically recorded or transmitted to a remote collection station.

Tipping gauges can also incorporate elements of weighing gauges whereby a strain gauge is fixed to the collection bucket so that the exact rainfall can be read at any moment. Each time the collector tips, the strain gauge (weight sensor) is re-zeroed to null out any drift.

To measure the *water equivalent* of frozen precipitation, a tipping bucket may be heated to melt any ice and snow that is caught in its funnel. Without a heating mechanism, the funnel often becomes clogged during a frozen precipitation event, and thus no precipitation can be measured. Many Automated Surface Observing System (ASOS) units use heated tipping buckets to measure precipitation.

Optical Rain Gauge

This type of gauge has a row of collection funnels. In an enclosed space below each is a laser diode and a photo transistor detector. When enough water is collected to make a single drop, it drops from the bottom, falling into the laser beam path. The sensor is set at right angles to the laser so that enough light is scattered to be detected as a sudden flash of lights. The flashes from these photodetectors are then read and transmitted or recorded.

Acoustic Rain Gauge

The acoustic disdrometer is an acoustic rain gauge. Also referred to as a hydrophone, it is able to sense the sound signatures for each drop size as rain strikes a water surface within the gauge. Since each sound signature is unique, it is possible to invert the underwater sound field to estimate the drop-size distribution within the rain. Selected moments of the drop-size distribution yield rainfall rate, rainfall accumulation, and other rainfall properties.

Importance of Rain Gauge

Simple but Effective Rain Gauge

Rain gauges are centuries-old tools, and some of those records of past rainfall patterns still exist. That data is invaluable to the researchers who study weather patterns and compare today's climate to that of the past.

Though some digital models exist today, the original design of the rain gauge hasn't altered much over the past few hundred years. There are a few different types of rain gauges, but most include a cylindrical cup and a funnel system. The design is simple yet effective. Usually, a funnel collects rain and then empties into a cylinder, which is inside of a larger container that collects excess rainwater. When it comes time to measure the rain, the rainfall totals in the larger container and the cylinder will give an exact reading of how much precipitation recently fell.

Monitoring for Droughts

One of the most important uses for rain gauges is to monitor droughts in areas reliant on agriculture, as well as cities that don't get much rain. For example, Los Angeles rain totals don't amount to much, compared to the rest of the country. The city frequently experiences droughts, which can be devastating for both home gardens and commercial agriculture in the surrounding areas. So, local farmers use rain gauges to measure Los Angeles rainfall. The information helps them know whether or not they'll have to dip into their reservoir or groundwater supply to help keep their crops alive, or prepare financially if one year's crop might not be as healthy as a different batch from a wetter year.

Meteorologists also use rain gauges to measure Los Angeles precipitation, as well as precipitation in other cities dealing with droughts. In addition to helping them create a weather forecast, they can also warn people when L.A.'s rainfall has been particularly low. They might advise citizens to cut back on using excess water in order not to exacerbate the drought.

Prepping for Disaster

A greater understanding of precipitation patterns can also help meteorologists prepare for weather-related natural disasters. For instance, they might recognize that recent rainfall patterns indicate a serious storm is on the way. Armed with this information, they can warn people about impending disaster, and advise them to protect their homes and avoid unnecessary travel. In this way, the data that comes from a rain gauge can help save buildings, crops and even lives.

References

- Barometer, encyclopedia: nationalgeographic.org, Retrieved 31 March, 2020
- Evangelista torricelli, the invention of the barometer". Juliantrubin.com. Archived from the original on 9 february 2010. Retrieved 2010-02-04
- Cengal, yunus a. And boles, michael a. (2014) thermodynamics: an engineering approach. Mc-graw-hill education. Isbn 978-0073398174
- Hurst, will; jessel, ella; waite, richard. "why architects need to wake up to the carbon emergency". Architects journal. Retrieved 2019-08-10
- Alvarez, d. A.; huckins,j. N.; petty, j.d.; jones-lepp, t.; stuer-lauridsen, f.; getting, d.t.; goddard, j.p.; gravel a. (2007). Comprehensive analytical chemistry: passive sampling techniques in environmental monitoring. Netherlands: elsevier. Pp. 171–197. Isbn 978-0-444-52225-2
- Salter, eddie. "active versus passive air sampling". Environmental technology. Labmate online. Retrieved 4 june 2018
- Noss, a.j.; oetting, i.; cuellar, r.l. (2005). "hunter self-monitoring by the isoseno-guarani in the bolivian chaco". Biodivers. Conserv. 14 (11): 2679–2693. Doi:10.1007/s10531-005-8401-2

3

Air Quality Monitoring

The quality of air is monitored in terms of the degree by which the air is clean and free from pollutants like smog, dust, smoke and other impurities of the air. Air quality index, air pollution forecasting and monitoring are some of the aspects that fall under its domain. The topics elaborated in this chapter will help in gaining a better perspective about air quality monitoring.

AIR QUALITY

“Air quality” refers to the condition of the air within our surrounding. Good air quality pertains to the degree which the air is clean, clear and free from pollutants such as smoke, dust and smog among other gaseous impurities in the air. Air quality is determined by assessing a variety of pollution indicators. Good air quality is a requirement for preserving the exquisite balance of life on earth for humans, plants, animals and natural resources. As such, human health, plants, animals and natural resources are threatened when pollution in the air reach high concentrations.

Poor air quality can affect or harm human health and/or the environment. Air quality can be degraded by natural or man-made sources. Natural sources include volcanic eruption, windstorm dust. Man-made source include pollution from moving vehicles, toxic gases from industries, coal powered plants, burning wood or other material in open air, landfills. Both these sources can seriously affect the overall air quality and can lead to severe health problems for humans.

Cause of Air Quality Degradation

Air quality is degraded by the day to day emissions from different air pollution sources. These include: stationary sources such as power plants, manufacturing facilities, smelters, motor machines, and burning wood and coal; mobile sources such as trucks, buses, planes and cars; and natural sources such as volcanic eruptions, and windstorm dust. The pollutants from these sources are released into the air which can affect the health of humans, animals and our environment. The quality of air depends on 3 factors: the amount of pollutants, the rate at which they are released in the atmosphere, and how long they are trapped in an area.

The air pollution sources emit a wide range of pollutants classified under different categories that include Sulfur Dioxide (SO_2), particulate matter, Hydrocarbons (HC) and volatile organic compounds (VOC), Lead, Carbon Dioxide (CO_2) and Carbon Monoxide (CO), Nitrogen Oxides (NOx), and Smog. In most regions, noticeable air quality degraders include smoke, dust, CO, NOx, SO_2, HC, and smog from factories, power plants, automobiles, and smelters especially around urban and industrial areas.

Many of these air pollutants occur as gases, but some are very tiny solid particles, such as dust, smoke or soot. Air pollution from various sources can affect indoor air as well. Removing the pollutants from indoor air can improve the quality of your indoor air. Indoor air pollutants include cigarette smoke, mould, dust mites, pet dander, formaldehyde, volatile organic compounds (VOCs) and radon gas.

If air pollutants are in an area with good airflow, they will mix with the air and quickly disperse. Air pollutants tend to remain in the air when when there are certain conditions like light winds or mountains that restrict the transport of these pollutants away from an area. When this happens, pollution concentrations can increase rapidly.

Improving Air Quality

With all those toxic pollutants in the atmosphere, it isn't easy to get fresh as it used to be. Due to increase in number of vehicles and industries on a large scale, the air quality has gone from bad to worse in lat few decades. Every year, millions of people across the globe die due to inhaling of toxic gases present in the atmosphere.

There are several measures which we can take to make air quality better not only for us but even for coming generations. If not done in proper and timely manner, the worsening air quality can cause severe harm to entire planetary ecosystem.

Dealing with Traffic and Vehicle Emissions

In a bid to improve air quality, dealing with automobile emission is the first choice for action. Beginning by reducing the numbers of automobile traffic and use of cleaner vehicles is the best course of action towards cutting down on emissions locally. Promoting public transport and proper urban planning can as well aid in reducing traffic numbers. Nowadays, journeys have become shorter and affordable due to introduction of fast electric trains and other forms of cheaper public transport.

By promoting public transport, it is easy to get people out of their vehicles which suggest that it's an obvious answer to reducing the number of cars on the road. Air pollution reduction can also be accomplished through speed reduction, promoting use of electric/hybrid cars, rerouting traffic, and introducing access restrictions. Through this, CO_2, lead, smoke, and particulate matter from road traffic emissions can considerably reduce thereby improving air quality. Besides use of cleaner vehicles and promotion of public transport, use of green fuels is a good alternative for improving air quality.

Dealing with Industrial Emissions

Factories and industrial manufacturing companies are another main concern when it comes to air pollution. Industries are the biggest air pollution sources as they emit high scores of pollutants into the air on daily basis. Due to the industrial emissions, the globe faces a great challenge of acid rain, ozone layer depletion, and global warming. Primary regulation policies by environmental agencies are required to regulate industrial emissions with an aim of large scale structural emission reduction from industrial facilities.

Control devices termed as "pollution eating nano-particles" can as well offer effective ways to reduce industrial air pollution by absorbing or destroying toxic emissions and contaminants from industries. It is realized through installation of particulate control devices in exhaust streams. Good examples include Titanium dioxide and ultraviolet light. Installed in an industrial facility, titanium dioxide reduces pollution by absorbing toxic emissions while ultraviolet light breaks up NOx gases and VOCs. Other control devices include electrostatic precipitators, dust cyclones, particulate scrubbers, and bag-houses. Switching to use of renewable energy is the most effective mechanism for improving air quality standards in industrial manufacturing.

Establishment of Policies and Clean Air Action Plans

Air quality improvement equally requires a string of directives, support, and initiatives geared towards emission reduction not only locally but internationally as well. Local, national and international environmental protection agencies need to work hand in hand to ensure the set air quality levels are met. It's a commitment that can only be achieved through an effective environmental action plan, strategies and policies directed towards addressing the underlying air pollution challenges. For instance, the Clean Air for Europe (CAFE) and the Environmental Protection Agency (EPA) programs have set standards for restricting air pollution and it has since resulted to a new directive for clean and quality air. Established about a decade ago, the directives have managed to set in place emission goals and standards for air quality.

Objectives are set whereby all sorts of air pollutants have to be regulated and cut down to a standard level and also managed within the scope of the challenges experienced. In challenging situations, deadlines for complying must be instituted and upheld. As such, creation of action plan directives similar to CAFE's, EPA's or any other devoted environmental protection agency can offer a wonderful advancement in the direction of improving air quality.

To work towards the attainment of improving the air quality situation, each state and nation without air regulation policies can make it a requirement to develop an action plan for cleaning the air within their jurisdiction. In addition to development of an action plan, implementation is essential as it defines the efforts that will improve and attempt to regulate the air pollution levels and to improve the air quality of the region.

Also through various environmental protection and clean air support programs, states and nation can establish a monitoring criteria for pollutants. One similar program is the Ambient Air Monitoring that has been adopted by various nations and states. Through the program, statistics can be collected regarding air quality by measuring samples so as to estimate the attainment of the set air quality standards. In this sense, pollution trends can be observed throughout the jurisdiction region to deter or reduce air pollution crisis. Other air quality monitoring programs can encompass Air Pollution Monitoring and Enhanced Ozone Monitoring.

Air pollution reduction efforts can also involve land-use, urban, and transportation planning strategies. The utilization of these strategies can assess the root causes of air pollution and to install measures for dealing with the challenges experienced.

Accordingly, environmental clean air policies and programs can be established to address issues regarding quality of air as a factor directly impacting on human health and the environment. The world, particularly regions burdened with the air pollution challenge, can thus improve air quality by adopting or establishing clean air action plans and support programs.

Indoor Air Quality

Indoor air quality (IAQ) is the air quality within and around buildings and structures. IAQ is known to affect the health, comfort and well-being of building occupants. Poor indoor air quality has been linked to sick building syndrome, reduced productivity and impaired learning in schools.

IAQ can be affected by gases (including carbon monoxide, radon, volatile organic compounds), particulates, microbial contaminants (mold, bacteria), or any mass or energy stressor that can induce adverse health conditions. Source control, filtration and the use of ventilation to dilute contaminants are the primary methods for improving indoor air quality in most buildings. Residential units can further improve indoor air quality by routine cleaning of carpets and area rugs.

Determination of IAQ involves the collection of air samples, monitoring human exposure to pollutants, collection of samples on building surfaces, and computer modelling of air flow inside buildings.

IAQ is part of indoor environmental quality (IEQ), which includes IAQ as well as other physical and psychological aspects of life indoors (e.g., lighting, visual quality, acoustics, and thermal comfort).

Indoor air pollution in developing nations is a major health hazard. A major source of indoor air pollution in developing countries is the burning of biomass (e.g. wood, charcoal, dung, or crop residue) for heating and cooking. The resulting exposure to high levels of particulate matter resulted in between 1.5 million and 2 million deaths in 2000.

Common Pollutants

Secondhand Smoke

Secondhand smoke is tobacco smoke which affects people other than the 'active' smoker. Second-hand tobacco smoke includes both a gaseous and a particulate phase, with particular hazards arising from levels of carbon monoxide and very small particulates (fine particular matter at especially PM2.5 size, and PM10) which get into the bronchioles and alveoles in the lung. The only certain method to improve indoor air quality as regards secondhand smoke is to eliminate smoking indoors.

Radon

Radon is an invisible, radioactive atomic gas that results from the radioactive decay of radium, which may be found in rock formations beneath buildings or in certain building materials themselves. Radon is probably the most pervasive serious hazard for indoor air in the United States and Europe, and is probably responsible for tens of thousands of deaths from lung cancer each year. There are relatively simple test kits for do-it-yourself radon gas testing, but if a home is for sale the testing must be done by a licensed person in some U.S. states. Radon gas enters buildings as a soil gas and is a heavy gas and thus will tend to accumulate at the lowest level. Radon may also be introduced into a building through drinking water particularly from bathroom showers. Building materials can be a rare source of radon, but little testing is carried out for stone, rock or tile products brought into building sites; radon accumulation is greatest for well insulated homes. The half life for radon is 3.8 days, indicating that once the source is removed, the hazard will be greatly reduced within a few weeks. Radon mitigation methods include sealing concrete slab floors, basement foundations, water drainage systems, or by increasing ventilation. They are usually cost effective and can greatly reduce or even eliminate the contamination and the associated health risks.

Radon is measured in picocuries per liter of air (pCi/L), a measurement of radioactivity. In the United States, the average indoor radon level is about 1.3 pCi/L. The average outdoor level is about 0.4 pCi/L. The U.S. Surgeon General and EPA recommend fixing homes with radon levels at or above 4 pCi/L. EPA also recommends that people think about fixing their homes for radon levels between 2 pCi/L and 4 pCi/L.

Molds and other Allergens

These biological chemicals can arise from a host of means, but there are two common classes: (a) moisture induced growth of mold colonies and (b) natural substances released into the air such as animal dander and plant pollen. Mold is always associated with moisture, and its growth can be inhibited by keeping humidity levels below 50%. Moisture buildup inside buildings may arise from water penetrating compromised areas of the building envelope or skin, from plumbing leaks, from condensation due to improper ventilation, or from ground moisture penetrating a building part. Even something as simple as drying clothes indoors on radiators can increase the risk of exposure to (amongst other things) Aspergillus – a highly dangerous mould that can be fatal for asthma sufferers and the elderly. In areas where cellulosic materials (paper and wood, including drywall) become moist and fail to dry within 48 hours, mold mildew can propagate and release allergenic spores into the air.

In many cases, if materials have failed to dry out several days after the suspected water event, mold growth is suspected within wall cavities even if it is not immediately visible. Through a mold investigation, which may include destructive inspection, one should be able to determine the presence or absence of mold. In a situation where there is visible mold and the indoor air quality may have been compromised, mold remediation may be needed. Mold testing and inspections should be carried out by an independent investigator to avoid any conflict of interest and to insure accurate results; free mold testing offered by remediation companies is not recommended.

There are some varieties of mold that contain toxic compounds (mycotoxins). However, exposure to hazardous levels of mycotoxin via inhalation is not possible in most cases, as toxins are produced by the fungal body and are not at significant levels in the released spores. The primary hazard of mold growth, as it relates to indoor air quality, comes from the allergenic properties of the spore cell wall. More serious than most allergenic properties is the ability of mold to trigger episodes in persons that already have asthma, a serious respiratory disease.

Carbon Monoxide

One of the most acutely toxic indoor air contaminants is carbon monoxide (CO), a colourless and odourless gas that is a by-product of incomplete combustion. Common sources of carbon monoxide are tobacco smoke, space heaters using fossil fuels, defective central heating furnaces and automobile exhaust. By depriving the brain of oxygen, high levels of carbon monoxide can lead to nausea, unconsciousness and death. According to the American Conference of Governmental Industrial Hygienists (ACGIH), the time-weighted average (TWA) limit for carbon monoxide (630–08–0) is 25 ppm.

Volatile Organic Compounds

Volatile organic compounds (VOCs) are emitted as gases from certain solids or liquids. VOCs include a variety of chemicals, some of which may have short- and long-term adverse health effects. Concentrations of many VOCs are consistently higher indoors (up to ten times higher) than outdoors. VOCs are emitted by a wide array of products numbering in the thousands. Examples include: paints and lacquers, paint strippers, cleaning supplies, pesticides, building materials and furnishings, office equipment such as copiers and printers, correction fluids and carbonless copy paper, graphics and craft materials including glues and adhesives, permanent markers, and photographic solutions.

Chlorinated drinking water releases chloroform when hot water is used in the home. Benzene is emitted from fuel stored in attached garages. Overheated cooking oils emit acrolein and formaldehyde. A meta-analysis of 77 surveys of VOCs in homes in the US found the top ten riskiest indoor air VOCs were acrolein, formaldehyde, benzene, hexachlorobutadiene, acetaldehyde, 1,3-butadiene, benzyl chloride, 1,4-dichlorobenzene, carbon tetrachloride, acrylonitrile, and vinyl chloride. These compounds exceeded health standards in most homes.

Organic chemicals are widely used as ingredients in household products. Paints, varnishes, and wax all contain organic solvents, as do many cleaning, disinfecting, cosmetic, degreasing, and hobby products. Fuels are made up of organic chemicals. All of these products can release organic compounds during usage, and, to some degree, when they are stored. Testing emissions from building materials used indoors has become increasingly common for floor coverings, paints, and many other important indoor building materials and finishes.

Several initiatives envisage to reduce indoor air contamination by limiting VOC emissions from products. There are regulations in France and in Germany, and numerous voluntary ecolabels and rating systems containing low VOC emissions criteria such as EMICODE, M1, Blue Angel and Indoor Air Comfort in Europe, as well as California Standard CDPH Section 01350 and several others in the USA. These initiatives changed the marketplace where an increasing number of low-emitting products has become available during the last decades.

At least 18 Microbial VOCs (MVOCs) have been characterised including 1-octen-3-ol, 3-methylfuran, 2-pentanol, 2-hexanone, 2-heptanone, 3-octanone, 3-octanol, 2-octen-1-ol, 1-octene, 2-pentanone, 2-nonanone, borneol, geosmin, 1-butanol, 3-methyl-1-butanol, 3-methyl-2-butanol, and thujopsene. The first of these compounds is called mushroom alcohol. The last four are products of *Stachybotrys chartarum*, which has been linked with sick building syndrome.

Legionella

Legionellosis or Legionnaire's Disease is caused by a waterborne bacterium *Legionella* that grows best in slow-moving or still, warm water. The primary route of exposure is through the creation of an aerosol effect, most commonly from evaporative cooling towers or showerheads. A common source of Legionella in commercial buildings is from poorly placed or maintained evaporative cooling towers, which often release water in an aerosol which may enter nearby ventilation intakes. Outbreaks in medical facilities and nursing homes, where patients are immuno-suppressed and immuno-weak, are the most commonly reported cases of Legionellosis. More than one case has involved outdoor fountains in public attractions. The presence of Legionella in commercial building water supplies is highly under-reported, as healthy people require heavy exposure to acquire infection.

Legionella testing typically involves collecting water samples and surface swabs from evaporative cooling basins, shower heads, faucets/taps, and other locations where warm water collects. The samples are then cultured and colony forming units (cfu) of Legionella are quantified as cfu/Liter.

Legionella is a parasite of protozoans such as amoeba, and thus requires conditions suitable for both organisms. The bacterium forms a biofilm which is resistant to chemical and antimicrobial treatments, including chlorine. Remediation for Legionella outbreaks in commercial buildings vary, but often include very hot water flushes (160 °F; 70 °C), sterilisation of standing water in evaporative cooling basins, replacement of shower heads, and in some cases flushes of heavy metal salts. Preventative measures include adjusting normal hot water levels to allow for 120 °F (50 °C) at the tap, evaluating facility design layout, removing faucet aerators, and periodic testing in suspect areas.

Other Bacteria

There are many bacteria of health significance found in indoor air and on indoor surfaces. The role of microbes in the indoor environment is increasingly studied using modern gene-based analysis of environmental samples. Currently efforts are under way to link microbial ecologists and indoor air scientists to forge new methods for analysis and to better interpret the results.

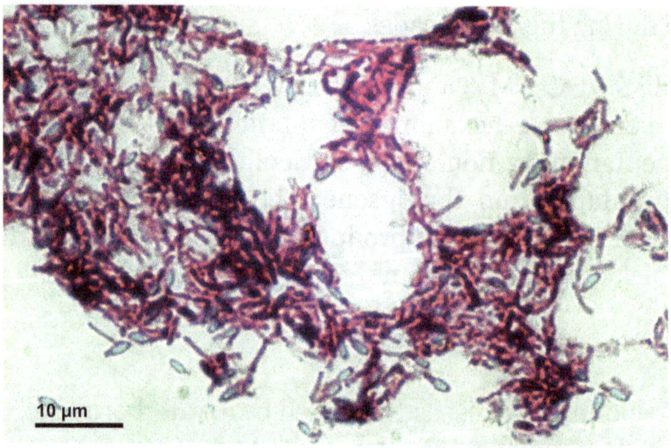

Bacteria (26 2 27) Airborne microbes.

"There are approximately ten times as many bacterial cells in the human flora as there are human cells in the body, with large numbers of bacteria on the skin and as gut flora." A large fraction of the bacteria found in indoor air and dust are shed from humans. Among the most important bacteria known to occur in indoor air are Mycobacterium tuberculosis, Staphylococcus aureus, Streptococcus pneumoniae.

Asbestos Fibers

Many common building materials used before 1975 contain asbestos, such as some floor tiles, ceiling tiles, shingles, fireproofing, heating systems, pipe wrap, taping muds, mastics, and other insulation materials. Normally, significant releases of asbestos fiber do not occur unless the building materials are disturbed, such as by cutting, sanding, drilling, or building remodelling. Removal of asbestos-containing materials is not always optimal because the fibers can be spread into the air during the removal process. A management program for intact asbestos-containing materials is often recommended instead.

When asbestos-containing material is damaged or disintegrates, microscopic fibers are dispersed into the air. Inhalation of asbestos fibers over long exposure times is associated with increased incidence of lung cancer, in particular the specific form mesothelioma. The risk of lung cancer from inhaling asbestos fibers is significantly greater to smokers, however there is no confirmed connection to damage caused by asbestosis . The symptoms of the disease do not usually appear until about 20 to 30 years after the first exposure to asbestos.

Asbestos is found in older homes and buildings, but occurs most commonly in schools, hospitals and industrial settings. Although all asbestos is hazardous, products that are friable, eg. sprayed coatings and insulation, pose a significantly higher hazard as they are more likely to release fibers to the air. The US Federal Government and some states have set standards for acceptable levels of asbestos fibers in indoor air. There are particularly stringent regulations applicable to schools.

Carbon Dioxide

Carbon dioxide (CO_2) is a relatively easy to measure surrogate for indoor pollutants emitted by humans, and correlates with human metabolic activity. Carbon dioxide at levels that are unusually high indoors may cause occupants to grow drowsy, to get headaches, or to function at lower activity levels. Outdoor CO_2 levels are usually 350–450 ppm whereas the maximum indoor CO_2 level considered acceptable is 1000 ppm. Humans are the main indoor source of carbon dioxide in most buildings. Indoor CO_2 levels are an indicator of the adequacy of outdoor air ventilation relative to indoor occupant density and metabolic activity.

To eliminate most complaints, the total indoor CO_2 level should be reduced to a difference of less than 600 ppm above outdoor levels. The National Institute for Occupational Safety and Health (NIOSH) considers that indoor air concentrations of carbon dioxide that exceed 1,000 ppm are a marker suggesting inadequate ventilation. The UK standards for schools say that carbon dioxide in all teaching and learning spaces, when measured at seated head height and averaged over the whole day should not exceed 1,500 ppm. The whole day refers to normal school hours (i.e. 9:00am to 3:30pm) and includes unoccupied periods such as lunch breaks. In Hong Kong, the EPD established indoor air quality objectives for office buildings and public places in which a carbon dioxide level below 1,000 ppm is considered to be good. European standards limit carbon dioxide to 3,500 ppm. OSHA limits carbon dioxide concentration in the workplace to 5,000 ppm for prolonged periods, and 35,000 ppm for 15 minutes. These higher limits are concerned with avoiding loss of consciousness (fainting), and do not address impaired cognitive performance and energy, which begin to occur at lower concentrations of carbon dioxide. Given the well established roles of oxygen sensing pathways in cancer and the acidosis independent role of carbon dioxide in modulating immune and inflammation linking pathways, it has been suggested that the effects of long-term indoor inspired elevated carbon dioxide levels on the modulation of carcinogenesis be investigated.

Carbon dioxide concentrations increase as a result of human occupancy, but lag in time behind cumulative occupancy and intake of fresh air. The lower the air exchange rate, the slower the buildup of carbon dioxide to quasi "steady state" concentrations on which the NIOSH and UK guidance are based. Therefore, measurements of carbon dioxide for purposes of assessing the adequacy of ventilation need to be made after an extended period of steady occupancy and ventilation – in schools at least 2 hours, and in offices at least 3 hours – for concentrations to be a reasonable indicator of ventilation adequacy. Portable instruments used to measure carbon dioxide should be calibrated frequently, and outdoor measurements used for calculations should be made close in time to indoor measurements. Corrections for temperature effects on measurements made outdoors may also be necessary.

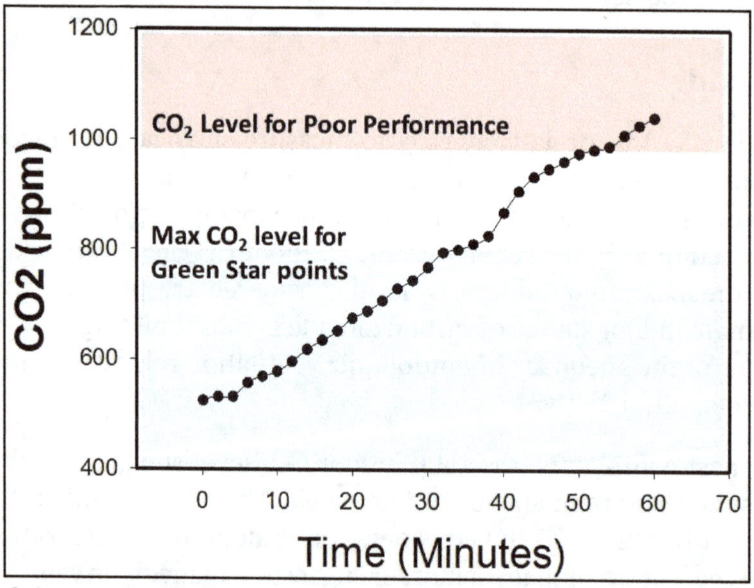

CO_2 levels in an enclosed office room can increase to over 1,000 ppm within 45 minutes.

Carbon dioxide concentrations in closed or confined rooms can increase to 1,000 ppm within 45 minutes of enclosure. For example, in a 3.5-by-4-metre (11 ft × 13 ft) sized office, atmospheric carbon dioxide increased from 500 ppm to over 1,000 ppm within 45 minutes of ventilation cessation and closure of windows and doors.

Ozone

Ozone is produced by ultraviolet light from the Sun hitting the Earth's atmosphere (especially in the ozone layer), lightning, certain high-voltage electric devices (such as air ionizers), and as a by-product of other types of pollution.

Ozone exists in greater concentrations at altitudes commonly flown by passenger jets. Reactions between ozone and onboard substances, including skin oils and cosmetics, can produce toxic chemicals as by-products. Ozone itself is also irritating to lung tissue and harmful to human health. Larger jets have ozone filters to reduce the cabin concentration to safer and more comfortable levels.

Outdoor air used for ventilation may have sufficient ozone to react with common indoor pollutants as well as skin oils and other common indoor air chemicals or surfaces. Particular concern is warranted when using "green" cleaning products based on citrus or terpene extracts, because these chemicals react very quickly with ozone to form toxic and irritating chemicals as well as fine and ultrafine particles. Ventilation with outdoor air containing elevated ozone concentrations may complicate remediation attempts.

Ozone is on the list of six criteria air pollutant list. The Clean Air Act of 1990 required the United States Environmental Protection Agency to set National Ambient Air Quality Standards (NAAQS) for six common indoor air pollutants harmful to human health. There are also multiple other organizations that have put forth air standards such as Occupational Safety and Health Administration (OSHA), National Institute for Occupational Safety and Health (NIOSH), and the World Health Organization (WHO). The OSHA standard for Ozone concentration within a space is 0.1 ppm. While the NAAQS and the EPA standard for ozone concentration is limited to 0.07 ppm. The type of ozone being regulated is ground-level ozone that is within the breathing range of most building occupants.

Particulates

Atmospheric particulate matter, also known as particulates, can be found indoors and can affect the health of occupants. Authorities have established standards for the maximum concentration of particulates to ensure indoor air quality.

Prompt Cognitive Deficits

In 2015, experimental studies reported the detection of significant episodic (situational) cognitive impairment from impurities in the air breathed by test subjects who were not informed about changes in the air quality. Researchers at the Harvard University and SUNY Upstate Medical University and Syracuse University measured the cognitive performance of 24 participants in three different controlled laboratory atmospheres that simulated those found in "conventional" and "green" buildings, as well as green buildings with enhanced ventilation. Performance was evaluated objectively using the widely used Strategic Management Simulation software simulation tool, which is a well-validated assessment test for executive decision-making in an unconstrained situation allowing initiative and improvisation. Significant deficits were observed in the performance scores achieved in increasing concentrations of either VOCs or carbon dioxide, while keeping other factors constant. The highest impurity levels reached are not uncommon in some classroom or office environments.

Effect of Indoor Plants

Houseplants together with the medium in which they are grown can reduce components of indoor air pollution, particularly volatile organic compounds (VOC) such as benzene, toluene, and xylene. Plants remove CO_2 and release oxygen and water, although the quantitative impact for house plants is small. Most of the effect is attributed to the growing medium alone, but even this effect has finite limits associated with the type and quantity of medium and the flow of air through the medium. The effect of house plants on VOC concentrations was investigated in one study, done in a static chamber, by NASA for possible use in space colonies. The results showed that the removal of the challenge chemicals was roughly equivalent to that provided by the ventilation that occurred in a very energy efficient dwelling with a very low ventilation rate, an air exchange rate of about 1/10 per hour. Therefore, air leakage in most homes, and in non-residential buildings too, will generally remove the chemicals faster than the researchers reported for the plants tested by NASA. The most effective household plants reportedly included aloe vera, English ivy, and Boston fern for removing chemicals and biological compounds.

Spider plants (Chlorophytum comosum) absorb some airborne contaminants.

Plants also appear to reduce airborne microbes and molds, and to increase humidity. However, the increased humidity can itself lead to increased levels of mold and even VOCs.

When carbon dioxide concentrations are elevated indoors relative to outdoor concentrations, it is only an indicator that ventilation is inadequate to remove metabolic products associated with human occupancy. Plants require carbon dioxide to grow and release oxygen when they consume carbon dioxide. Akira Tani and C. Nicholas Hewitt found "Longer-term fumigation results revealed that the total uptake amounts were 30–100 times as much as the amounts dissolved in the leaf, suggesting that volatile organic carbons are metabolized in the leaf and/or translocated through the petiole." It is worth noting the researchers sealed the plants in Teflon bags. "No VOC loss was detected from the bag when the plants were absent. However, when the plants were in the bag, the levels of aldehydes and ketones both decreased slowly but continuously, indicating removal by the plants." Studies done in sealed bags do not faithfully reproduce the conditions in the indoor environments of interest. Dynamic conditions with outdoor air ventilation and the processes related to the surfaces of the building itself and its contents as well as the occupants need to be studied.

While results do indicate house plants may be effective at removing some VOCs from air supplies, a review of studies between 1989 and 2006 on the performance of houseplants as air cleaners, presented at the Healthy Buildings 2009 conference in Syracuse, New York, concluded "indoor plants have little, if any, benefit for removing indoor air of VOC in residential and commercial buildings." This conclusion was based on a trial involving an unknown quantity of indoor plants kept in an uncontrolled ventilated air environment of an arbitrary office building in Arlington, Virginia.

Since extremely high humidity is associated with increased mold growth, allergic responses, and respiratory responses, the presence of additional moisture from houseplants may not be desirable in all indoor settings if watering is done inappropriately.

HVAC Design

Environmentally sustainable design concepts also include aspects related to the commercial and residential heating, ventilation and air-conditioning (HVAC) industry. Among several considerations, one of the topics attended to is the issue of indoor air quality throughout the design and construction stages of a building's life.

One technique to reduce energy consumption while maintaining adequate air quality, is demand-controlled ventilation. Instead of setting throughput at a fixed air replacement rate, carbon dioxide sensors are used to control the rate dynamically, based on the emissions of actual building occupants.

For the past several years, there have been many debates among indoor air quality specialists about the proper definition of indoor air quality and specifically what constitutes "acceptable" indoor air quality.

One way of quantitatively ensuring the health of indoor air is by the frequency of effective turnover of interior air by replacement with outside air. In the UK, for example, classrooms are required to have 2.5 outdoor air changes per hour. In halls, gym, dining, and physiotherapy spaces, the ventilation should be sufficient to limit carbon dioxide to 1,500 ppm. In the USA, and according to ASHRAE Standards, ventilation in classrooms is based on the amount of outdoor air per occupant plus the amount of outdoor air per unit of floor area, not air changes per hour. Since carbon dioxide indoors comes from occupants and outdoor air, the adequacy of ventilation per occupant is indicated by the concentration indoors minus the concentration outdoors. The value of 615 ppm above the outdoor concentration indicates approximately 15 cubic feet per minute of outdoor air per adult occupant doing sedentary office work where outdoor air contains 385 ppm, the current global average atmospheric CO_2 concentration. In classrooms, the requirements in the ASHRAE standard 62.1, Ventilation for Acceptable Indoor Air Quality, would typically result in about 3 air changes per hour, depending on the occupant density. Of course the occupants are not the only source of pollutants, so outdoor air ventilation may need to be higher when unusual or strong sources of pollution exist indoors. When outdoor air is polluted, then bringing in more outdoor air can actually worsen the overall quality of the indoor air and exacerbate some occupant symptoms related to outdoor air pollution. Generally, outdoor country air is better than indoor city air. Exhaust gas leakages can occur from furnace metal exhaust pipes that lead to the chimney when there are leaks in the pipe and the pipe gas flow area diameter has been reduced.

The use of air filters can trap some of the air pollutants. The Department of Energy's Energy Efficiency and Renewable Energy topic suggests that "Air Filtration should have a Minimum Efficiency Reporting Value (MERV) of 13 as determined by ASHRAE 52.2-1999." Air filters are used to reduce the amount of dust that reaches the wet coils. Dust can serve as food to grow molds on the wet coils and ducts and can reduce the efficiency of the coils.

Moisture management and humidity control requires operating HVAC systems as designed. Moisture management and humidity control may conflict with efforts to try to optimize the operation to conserve energy. For example, moisture management and humidity control requires systems to be set to supply make-up air at lower temperatures (design levels), instead of the higher temperatures sometimes used to conserve energy in cooling-dominated climate conditions. However, for most of the US and many parts of Europe and Japan, during the majority of hours of the year, outdoor air temperatures are cool enough that the air does not need further cooling to provide thermal comfort indoors. However, high humidity outdoors creates the need for careful attention to humidity levels indoors. High humidities give rise to mold growth and moisture indoors is associated with a higher prevalence of occupant respiratory problems.

The "dew point temperature" is an absolute measure of the moisture in air. Some facilities are being designed with the design dew points in the lower 50s °F, and some in the upper and lower 40s °F. Some facilities are being designed using desiccant wheels with gas-fired heaters to dry out the wheel enough to get the required dew points. On those systems, after the moisture is removed from the make-up air, a cooling coil is used to lower the temperature to the desired level.

Commercial buildings, and sometimes residential, are often kept under slightly positive air pressure relative to the outdoors to reduce infiltration. Limiting infiltration helps with moisture management and humidity control.

Dilution of indoor pollutants with outdoor air is effective to the extent that outdoor air is free of harmful pollutants. Ozone in outdoor air occurs indoors at reduced concentrations because ozone is highly reactive with many chemicals found indoors. The products of the reactions between ozone and many common indoor pollutants include organic compounds that may be more odorous, irritating, or toxic than those from which they are formed. These products of ozone chemistry include formaldehyde, higher molecular weight aldehydes, acidic aerosols, and fine and ultrafine particles, among others. The higher the outdoor ventilation rate, the higher the indoor ozone concentration and the more likely the reactions will occur, but even at low levels, the reactions will take place. This suggests that ozone should be removed from ventilation air, especially in areas where outdoor ozone levels are frequently high. Recent research has shown that mortality and morbidity increase in the general population during periods of higher outdoor ozone and that the threshold for this effect is around 20 parts per billion (ppb).

AIR QUALITY INDEX

Air quality index (AQI) is a numerical scale used for reporting day to day air quality with regard to human health and the environment. The daily results of the index are used to convey to the public an estimate of air pollution level. An increase in air quality index signifies increased air pollution and severe threats to human health. In most cases, AQI indicates how clear or polluted the air in our surrounding is, and the associated health risks it might present. The AQI centers on the health effects that may be experienced within a few days or hours after breathing polluted air.

AQI calculations focus on major air pollutants including: particulate matter, ground-level ozone, sulfur dioxide (SO_2), nitrogen dioxide (NO_2), and carbon monoxide (CO). Particulate matter and ozone pollutants pose the highest risks to human health and the environment. For each of these air pollutant categories, different countries have their own established air quality indices in relation to other nationally set air quality standards for public health protection.

Air Quality Determination

On an hourly basis, the concentration of each pollutant in the air is measured and converted into a number running from zero upwards by using a standard index or scale. Calculated number for every pollutant is termed as a sub-index. The highest sub-index for any given hour is recorded as the AQI for that hour. In simple terms, AQI is like a yardstick that ranges from zero to five hundred (0-500). The index is a relative scale, meaning, the lower the index the better the quality of air and the lesser the health concern, and vice versa. The concentration of each pollutant varies, therefore, AQI values are grouped into ranges assigned to a standardized public health warnings and color code.

For instance, an AQI value of 0 to 50 means good air quality with limited possibility of affecting public and environmental health. On the other hand, an AQI value of 300 to 500 represents hazardous air quality with greater potential to affect public and environment health. The commonly accepted value is 100 as it corresponds to the generally approved air quality standards as set to safeguard public health. AQI levels below 100 are highly satisfactory while values beyond 100 are regarded to harm human health. As the AQI values get higher, it even poses more serious health concerns.

During days that the AQI is recorded to be elevated, the institution of public health might: offer advice to sensitive groups such as the children, those with respiratory problems, and the elderly to keep away from outdoor activities; take action to improve air quality by introducing measures for reducing emissions; or recommend the use of air pollution masks, especially in severe cases of air pollution.

Formula for Calculating Air Quality Index

Air quality index is calculated by a linear function by determining the concentration of the pollutant. The equation below is used to compute AQI.

$$I = \frac{I_{high} - I_{low}}{C_{high} - C_{low}} (C - C_{low}) + I_{low}$$

Where:

I = The (Air Quality) index,

C = The pollutant concentration,

$C_{\{low\}}$ = The concentration breakpoint that is ≤ C,

$C_{\{high\}}$ = The concentration breakpoint that is ≥ C,

$I_{\{low\}}$ = The index breakpoint corresponding to $C_{\{low\}}$,

$I_{\{high\}}$ = The index breakpoint corresponding to $C_{\{high\}}$.

Air Quality Index Categories

The AQI is divided in six categories and each category is meant to correspond to different health concern levels. Below is an explanation of the categories and their meanings.

- 0 – 50 indicates "Good" AQI. At this level the quality of air is deemed to be satisfactory, and air pollution poses little or no risk.
- 51 – 100 indicate "Moderate" AQI. This means acceptable Air quality. However, some pollutants might arouse modest health concern for a limited number of people. For instance, persons who are remarkably sensitive to ozone may experience respiratory symptoms.
- 101 – 150 indicate "Unhealthy for Sensitive Groups" AQI. This category may not be able to affect the general health of the public. However, children, older adults, and persons with lung disease are at a greater risk from ozone exposure. Older children, adults and people with lung and heart disease are at greater risk from exposure to particulate matter.
- 151 – 200 indicate "Unhealthy" AQI. In this category, every person might experience some adverse health effects. Sensitive group members may experience more serious effects. People with heart or lung disease, older adults and children should cut back or reschedule strenuous activities.
- 201 – 300 indicate "Very Unhealthy" AQI. This would issue a health alert to mean that everybody may experience very serious health implications. People with heart or lung disease, older adults and children should significantly cut back or reschedule strenuous activities.
- Greater than 300 indicate "Hazardous" AQI. Air quality at this level is life-threatening and would issue warnings of emergency conditions for the entire population.

At a glance, the table below provides the Air Quality Index (AQI) categories.

Air Quality Index (AQI) Values	Levels of Health Concern	Colors
When the AQI is in this Range:	Air Quality Conditions are:	As Symbolized by this Color:
0-50	Good	Green
51-100	Moderate	Yellow
101-150	Unhealthy for Sensitive Groups	Orange
151 to 200	Unhealthy	Red
201 to 300	Very Unhealthy	Purple
301 to 500	Hazardous	Maroon

The AQI is calculated for four major air pollutants regulated by the Clean Air Act: ground-level ozone, particle pollution, carbon monoxide, and sulfur dioxide. You need to take following simple steps to avoid exposure to unhealthy air:

Prolonged Exertion: Prolonged exertion is any outdoor activity that you do intermittently for several hours and may cause you to breather slightly faster than normal. When air is unhealthy outside, you can reduce intake of unhealthy air by reducing how much time you spend on this type of activity.

Heavy Exertion: Heavy exertion means intense outdoor activities that cause you to breathe hard. When air quality is bad outside, you can protect your health by reducing the amount of time you spend on this activity or by substituting it with less intense activity.

AIR POLLUTION

Air pollution occurs when harmful or excessive quantities of substances including gases (CO_2, CO, SO_2, NO, CH_4, CFC, Radon, etc.), particles (both organic and inorganic), and biological molecules are introduced into Earth's atmosphere. It may cause diseases, allergies and even death to humans; it may also cause harm to other living organisms such as animals and food crops, and may damage the natural or built environment. Both human activity and natural processes can generate air pollution.

Indoor air pollution and poor urban air quality are listed as two of the world's worst toxic pollution problems in the 2008 Blacksmith Institute World's Worst Polluted Places report. Outdoor air pollution alone causes 2.1 to 4.21 million premature deaths annually. According to the 2014 World Health Organization report, air pollution in 2012 caused the deaths of around 7 million people worldwide, an estimate roughly echoed by the International Energy Agency. According to the Journal Elements, humankind generates up to 100 times as much CO_2 as venting from volcanoes each year.

Air pollution from a coking oven.

Pollutants

Before flue-gas desulphurization was installed, the emissions from this power plant in New Mexico contained excessive amounts of sulphur dioxide.

An air pollutant is a material in the air that can have adverse effects on humans and the ecosystem. The substance can be solid particles, liquid droplets, or gases. A pollutant can be of natural origin or man-made. Pollutants are classified as primary or secondary. Primary pollutants are usually produced by processes such as ash from a volcanic eruption. Other examples include carbon monoxide gas from motor vehicle exhausts or sulphur dioxide released from factories. Secondary pollutants are not emitted directly. Rather, they form in the air when primary pollutants react or interact. Ground level ozone is a prominent example of secondary pollutants. Some pollutants may be both primary and secondary: they are both emitted directly and formed from other primary pollutants.

Pollutants emitted into the atmosphere by human activity include:

- Carbon dioxide (CO_2): Because of its role as a greenhouse gas it has been described as "the leading pollutant" and "the worst climate pollution". Carbon dioxide is a natural component of the atmosphere, essential for plant life and given off by the human respiratory system. This question of terminology has practical effects, for example as determining whether the U.S. Clean Air Act is deemed to regulate CO_2 emissions. CO_2 currently forms about 410 parts per million (ppm) of earth's atmosphere, compared to about 280 ppm in pre-industrial times, and billions of metric tons of CO_2 are emitted annually by burning of fossil fuels. CO_2 increase in earth's atmosphere has been accelerating.

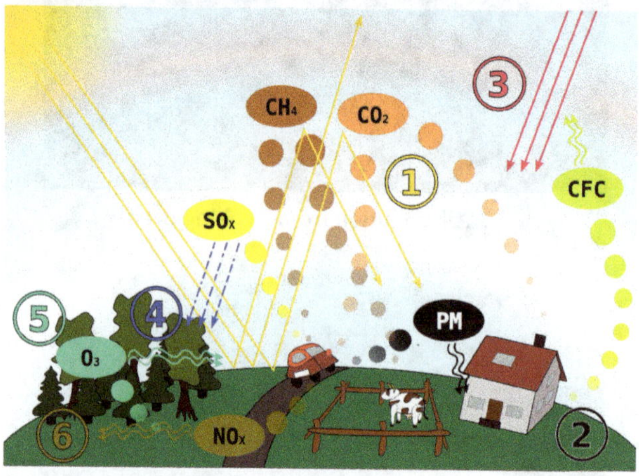

Schematic drawing, causes and effects of air pollution: (1) greenhouse effect, (2) particulate contamination, (3) increased UV radiation, (4) acid rain, (5) increased ground level ozone concentration, (6) increased levels of nitrogen oxides.

- Sulfur oxides (SO_x): Particularly sulphur dioxide, a chemical compound with the formula SO_2. SO_2 is produced by volcanoes and in various industrial processes. Coal and petroleum often contain sulphur compounds, and their combustion generates sulphur dioxide. Further oxidation of SO_2, usually in the presence of a catalyst such as NO_2, forms H_2SO_4, and thus acid rain. This is one of the causes for concern over the environmental impact of the use of these fuels as power sources.

Thermal oxidisers are air pollution abatement options for hazardous air pollutants (HAPs), volatile organic compounds (VOCs), and odorous emissions.

- Nitrogen oxides (NO_x): Nitrogen oxides, particularly nitrogen dioxide, are expelled from high temperature combustion, and are also produced during thunderstorms by electric discharge. They can be seen as a brown haze dome above or a plume downwind of cities. Nitrogen dioxide is a chemical compound with the formula NO_2. It is one of several nitrogen oxides. One of the most prominent air pollutants, this reddish-brown toxic gas has a characteristic sharp, biting odor.

- Carbon monoxide (CO): CO is a colorless, odorless, toxic yet non-irritating gas. It is a product of combustion of fuel such as natural gas, coal or wood. Vehicular exhaust contributes to the majority of carbon monoxide let into our atmosphere. It creates a smog type formation in the air that has been linked to many lung diseases and disruptions to the natural environment and animals. In 2013, more than half of the carbon monoxide emitted into our atmosphere was from vehicle traffic and burning one gallon of gas will often emit over 20 pounds of carbon monoxide into the air.

- Volatile organic compounds (VOC): VOCs are a well-known outdoor air pollutant. They are categorized as either methane (CH_4) or non-methane (NMVOCs). Methane is an extremely efficient greenhouse gas which contributes to enhanced global warming. Other hydrocarbon VOCs are also significant greenhouse gases because of their role in creating ozone and prolonging the life of methane in the atmosphere. This effect varies depending on local air quality. The aromatic NMVOCs benzene, toluene and xylene are suspected carcinogens and may lead to leukemia with prolonged exposure. 1,3-butadiene is another dangerous compound often associated with industrial use.

- Particulate matter/particles, alternatively referred to as particulate matter (PM), atmospheric particulate matter, or fine particles, are tiny particles of solid or liquid suspended in a gas. In contrast, aerosol refers to combined particles and gas. Some particulates occur naturally, originating from volcanoes, dust storms, forest and grassland fires, living vegetation, and sea spray. Human activities, such as the burning of fossil fuels in vehicles, power plants and various industrial processes also generate significant amounts of aerosols. Averaged worldwide, anthropogenic aerosols—those made by human activities—currently account for approximately 10 percent of our atmosphere. Increased levels of fine particles in the air are linked to health hazards such as heart disease, altered lung function and lung cancer. Particulates are related to respiratory infections and can be particularly harmful to those already suffering from conditions like asthma.

- Persistent free radicals connected to airborne fine particles are linked to cardiopulmonary disease.

- Toxic metals, such as lead and mercury, especially their compounds.

- Chlorofluorocarbons (CFCs): Harmful to the ozone layer; emitted from products are currently banned from use. These are gases which are released from air conditioners, refrigerators, aerosol sprays, etc. On release into the air, CFCs rise to the stratosphere. Here they come in contact with other gases and damage the ozone layer. This allows harmful ultraviolet rays to reach the earth's surface. This can lead to skin cancer, eye disease and can even cause damage to plants.
- Ammonia: Emitted mainly by agricultural waste. Ammonia is a compound with the formula NH_3. It is normally encountered as a gas with a characteristic pungent odor. Ammonia contributes significantly to the nutritional needs of terrestrial organisms by serving as a precursor to foodstuffs and fertilizers. Ammonia, either directly or indirectly, is also a building block for the synthesis of many pharmaceuticals. Although in wide use, ammonia is both caustic and hazardous. In the atmosphere, ammonia reacts with oxides of nitrogen and sulphur to form secondary particles.
- Odors: Such as from garbage, sewage, and industrial processes.
- Radioactive pollutants: Produced by nuclear explosions, nuclear events, war explosives, and natural processes such as the radioactive decay of radon.

Secondary pollutants include:

- Particulates created from gaseous primary pollutants and compounds in photochemical smog. Smog is a kind of air pollution. Classic smog results from large amounts of coal burning in an area caused by a mixture of smoke and sulphur dioxide. Modern smog does not usually come from coal but from vehicular and industrial emissions that are acted on in the atmosphere by ultraviolet light from the sun to form secondary pollutants that also combine with the primary emissions to form photochemical smog.
- Ground level ozone (O_3) formed from NO_x and VOCs. Ozone (O_3) is a key constituent of the troposphere. It is also an important constituent of certain regions of the stratosphere commonly known as the Ozone layer. Photochemical and chemical reactions involving it drive many of the chemical processes that occur in the atmosphere by day and by night. At abnormally high concentrations brought about by human activities (largely the combustion of fossil fuel), it is a pollutant and a constituent of smog.
- Peroxyacetyl nitrate ($C_2H_3NO_5$) – similarly formed from NO_x and VOCs.

Minor air pollutants include:

- A large number of minor hazardous air pollutants. Some of these are regulated in USA under the Clean Air Act and in Europe under the Air Framework Directive.
- A variety of persistent organic pollutants, which can attach to particulates.

Persistent organic pollutants (POPs) are organic compounds that are resistant to environmental degradation through chemical, biological, and photolytic processes. Because of this, they have been observed to persist in the environment, to be capable of long-range transport, bioaccumulate in human and animal tissue, biomagnify in food chains, and to have potentially significant impacts on human health and the environment.

There are various locations, activities or factors which are responsible for releasing pollutants into the atmosphere. These sources can be classified into two major categories.

Anthropogenic (Man-made) Sources

Controlled burning of a field outside of Statesboro, Georgia in preparation for spring planting.

Smoking of fish over an open fire.

These are mostly related to the burning of multiple types of fuel:

- Stationary sources include smoke stacks of fossil fuel power stations, manufacturing facilities (factories) and waste incinerators, as well as furnaces and other types of fuel-burning heating devices. In developing and poor countries, traditional biomass burning is the major source of air pollutants; traditional biomass includes wood, crop waste and dung.
- Mobile sources include motor vehicles, marine vessels, and aircraft.

- Controlled burn practices in agriculture and forest management. Controlled or prescribed burning is a technique sometimes used in forest management, farming, prairie restoration or greenhouse gas abatement. Fire is a natural part of both forest and grassland ecology and controlled fire can be a tool for foresters. Controlled burning stimulates the germination of some desirable forest trees, thus renewing the forest.

There are also sources from processes other than combustion:

- Fumes from paint, hair spray, varnish, aerosol sprays and other solvents. These can be substantial; emissions from these sources was estimated to account for almost half of pollution from volatile organic compounds in the Los Angeles basin in the 2010s.

- Waste deposition in landfills, which generate methane. Methane is highly flammable and may form explosive mixtures with air. Methane is also an asphyxiant and may displace oxygen in an enclosed space. Asphyxia or suffocation may result if the oxygen concentration is reduced to below 19.5% by displacement.

- Military resources, such as nuclear weapons, toxic gases, germ warfare and rocketry.

- Fertilized farmland may be a major source of nitrogen oxides.

Natural Sources

- Dust from natural sources, usually large areas of land with little or no vegetation.

- Methane, emitted by the digestion of food by animals, for example cattle.

- Radon gas from radioactive decay within the Earth's crust. Radon is a colorless, odorless, naturally occurring, radioactive noble gas that is formed from the decay of radium. It is considered to be a health hazard. Radon gas from natural sources can accumulate in buildings, especially in confined areas such as the basement and it is the second most frequent cause of lung cancer, after cigarette smoking.

Dust storm approaching Stratford, Texas.

- Smoke and carbon monoxide from wildfires. During periods of actives wildfires, smoke from uncontrolled biomass combustion can make up almost 75% of all air pollution by concentration.
- Vegetation, in some regions, emits environmentally significant amounts of Volatile organic compounds (VOCs) on warmer days. These VOCs react with primary anthropogenic pollutants—specifically, NO_x, SO_2, and anthropogenic organic carbon compounds — to produce a seasonal haze of secondary pollutants. Black gum, poplar, oak and willow are some examples of vegetation that can produce abundant VOCs. The VOC production from these species result in ozone levels up to eight times higher than the low-impact tree species.
- Volcanic activity, which produces sulphur, chlorine, and ash particulates.

Emission Factors

Beijing air on a 2005-day after rain (left) and a smoggy day (right).

Air pollutant emission factors are reported representative values that attempt to relate the quantity of a pollutant released to the ambient air with an activity associated with the release of that pollutant. These factors are usually expressed as the weight of pollutant divided by a unit weight, volume, distance, or duration of the activity emitting the pollutant (e.g., kilograms of particulate emitted per tonne of coal burned). Such factors facilitate estimation of emissions from various sources of air pollution. In most cases, these factors are simply averages of all available data of acceptable quality, and are generally assumed to be representative of long-term averages.

There are 12 compounds in the list of persistent organic pollutants. Dioxins and furans are two of them and intentionally created by combustion of organics, like open burning of plastics. These compounds are also endocrine disruptors and can mutate the human genes.

The United States Environmental Protection Agency has published a compilation of air pollutant emission factors for a wide range of industrial sources. The United Kingdom, Australia, Canada and many other countries have published similar compilations, as well as the European Environment Agency.

Exposure

Up to 30 % of Europeans living in cities are exposed to air pollutant levels exceeding EU air quality standards.
And around 98 % of Europeans living in cities are exposed to levels of air pollutants deemed damaging to health by the World Health Organization's more stringent guidelines.

Air pollution risk is a function of the hazard of the pollutant and the exposure to that pollutant. Air pollution exposure can be expressed for an individual, for certain groups (e.g. neighborhoods or children living in a country), or for entire populations. For example, one may want to calculate the exposure to a hazardous air pollutant for a geographic area, which includes the various microenvironments and age groups. This can be calculated as an inhalation exposure. This would account for daily exposure in various settings (e.g. different indoor micro-environments and outdoor locations). The exposure needs to include different age and other demographic groups, especially infants, children, pregnant women and other sensitive subpopulations. The exposure to an air pollutant must integrate the concentrations of the air pollutant with respect to the time spent in each setting and the respective inhalation rates for each subgroup for each specific time that the subgroup is in the setting and engaged in particular activities (playing, cooking, reading, working, spending time in traffic, etc). For example, a small child's inhalation rate will be less than that of an adult. A child engaged in vigorous exercise will have a higher respiration rate than the same child in a sedentary activity. The daily exposure, then, needs to reflect the time spent in each micro-environmental setting and the type of activities in these settings. The air pollutant concentration in each microactivity/microenvironmental setting is summed to indicate the exposure. For some pollutants such as black carbon, traffic related exposures may dominate total exposure despite short exposure times since high concentrations coincide with proximity to major roads or participation to (motorized) traffic. A large portion of total daily exposure occurs as short peaks of high concentrations, but it remains unclear how to define peaks and determine their frequency and health impact.

Health Effects

In 2012, air pollution caused premature deaths on average of 1 year in Europe, and was a significant risk factor for a number of pollution-related diseases, including respiratory infections, heart disease, COPD, stroke and lung cancer. The health effects caused by air pollution may include difficulty in breathing, wheezing, coughing, asthma and worsening of existing respiratory and cardiac conditions. These effects can result in increased medication use, increased doctor or emergency department visits, more hospital admissions and premature death. The human health effects of poor air quality are far reaching, but principally affect the body's respiratory system and the cardiovascular system. Individual reactions to air pollutants depend on the type of pollutant a person is exposed to, the degree of exposure, and the individual's health status and genetics. The most common sources of air pollution include particulates, ozone, nitrogen dioxide, and sulphur dioxide. Children aged less than five years that live in developing countries are the most vulnerable population in terms of total deaths attributable to indoor and outdoor air pollution.

Mortality

The World Health Organization estimated in 2014 that every year air pollution causes the premature death of some 7 million people worldwide. Studies published in March 2019 indicated that the number may be around 8.8 million.

India has the highest death rate due to air pollution. India also has more deaths from asthma than any other nation according to the World Health Organization. In December 2013 air pollution was estimated to kill 500,000 people in China each year. There is a positive correlation between pneumonia-related deaths and air pollution from motor vehicle emissions.

Annual premature European deaths caused by air pollution are estimated at 430,000-800,000 An important cause of these deaths is nitrogen dioxide and other nitrogen oxides (NOx) emitted by road vehicles. In a 2015 consultation document the UK government disclosed that nitrogen dioxide is responsible for 23,500 premature UK deaths per annum. Across the European Union, air pollution is estimated to reduce life expectancy by almost nine months. Causes of deaths include strokes, heart disease, COPD, lung cancer, and lung infections.

Urban outdoor air pollution is estimated to cause 1.3 million deaths worldwide per year. Children are particularly at risk due to the immaturity of their respiratory organ systems.

The US EPA estimated in 2004 that a proposed set of changes in diesel engine technology (*Tier 2*) could result in 12,000 fewer *premature mortalities*, 15,000 fewer heart attacks, 6,000 fewer emergency department visits by children with asthma, and 8,900 fewer respiratory-related hospital admissions each year in the United States.

The US EPA has estimated that limiting ground-level ozone concentration to 65 parts per billion, would avert 1,700 to 5,100 premature deaths nationwide in 2020 compared with the 75-ppb standard. The agency projected the more protective standard would also prevent an additional 26,000 cases of aggravated asthma, and more than a million cases of missed work or school. Following this assessment, the EPA acted to protect public health by lowering the National Ambient Air Quality Standards (NAAQS) for ground-level ozone to 70 parts per billion (ppb).

A new economic study of the health impacts and associated costs of air pollution in the Los Angeles Basin and San Joaquin Valley of Southern California shows that more than 3,800 people die prematurely (approximately 14 years earlier than normal) each year because air pollution levels violate federal standards. The number of annual premature deaths is considerably higher than the fatalities related to auto collisions in the same area, which average fewer than 2,000 per year.

Diesel exhaust (DE) is a major contributor to combustion-derived particulate matter air pollution. In several human experimental studies, using a well-validated exposure chamber setup, DE has been linked to acute vascular dysfunction and increased thrombus formation.

The mechanisms linking air pollution to increased cardiovascular mortality are uncertain, but probably include pulmonary and systemic inflammation.

Cardiovascular Disease

A 2007 review of evidence found ambient air pollution exposure is a risk factor correlating with increased total mortality from cardiovascular events (range: 12% to 14% per 10 microg/m^3 increase).

Air pollution is also emerging as a risk factor for stroke, particularly in developing countries where pollutant levels are highest. A 2007 study found that in women, air pollution is not associated with hemorrhagic but with ischemic stroke. Air pollution was also found to be associated with increased incidence and mortality from coronary stroke in a cohort study in 2011. Associations are believed to be causal and effects may be mediated by vasoconstriction, low-grade inflammation and atherosclerosis Other mechanisms such as autonomic nervous system imbalance have also been suggested.

Lung Disease

Research has demonstrated increased risk of developing asthma and COPD from increased exposure to traffic-related air pollution. Additionally, air pollution has been associated with increased hospitalization and mortality from asthma and COPD. Chronic obstructive pulmonary disease (COPD) includes diseases such as chronic bronchitis and emphysema.

A study conducted in 1960–1961 in the wake of the Great Smog of 1952 compared 293 London residents with 477 residents of Gloucester, Peterborough, and Norwich, three towns with low reported death rates from chronic bronchitis. All subjects were male postal truck drivers aged 40 to 59. Compared to the subjects from the outlying towns, the London subjects exhibited more severe respiratory symptoms (including cough, phlegm, and dyspnea), reduced lung function (FEV_1 and peak flow rate), and increased sputum production and purulence. The differences were more pronounced for subjects aged 50 to 59. The study controlled for age and smoking habits, so concluded that air pollution was the most likely cause of the observed differences. More recent studies have shown that air pollution exposure from traffic reduces lung function development in children and lung function may be compromised by air pollution even at low concentrations. Air pollution exposure also cause lung cancer in non smokers.

It is believed that much like cystic fibrosis, by living in a more urban environment serious health hazards become more apparent. Studies have shown that in urban areas patients suffer mucus hypersecretion, lower levels of lung function, and more self-diagnosis of chronic bronchitis and emphysema.

Cancer (Lung Cancer)

A review of evidence regarding whether ambient air pollution exposure is a risk factor for cancer in 2007 found solid data to conclude that long-term exposure to PM2.5 (fine particulates) increases the overall risk of non-accidental mortality by 6% per a 10 microg/m^3 increase. Exposure to PM2.5 was also associated with an increased risk of mortality from lung cancer (range: 15% to 21% per 10 microg/m^3 increase) and total cardiovascular mortality (range: 12% to 14% per a 10 microg/m^3 increase). The review further noted that living close to busy traffic appears to be associated with elevated risks of these three outcomes – increase in lung cancer deaths, cardiovascular deaths, and overall non-accidental deaths. The reviewers also found suggestive evidence that exposure to PM2.5 is positively associated with mortality from coronary heart diseases and exposure to SO_2 increases mortality from lung cancer, but the data was insufficient to provide solid conclusions. Another investigation showed that higher activity level increases deposition fraction of aerosol particles in human lung and recommended avoiding heavy activities like running in outdoor space at polluted areas.

Unprotected exposure to PM2.5 air pollution can be equivalent to smoking multiple cigarettes per day, potentially increasing the risk of Cancer, which is mainly the result of environmental factors.

In 2011, a large Danish epidemiological study found an increased risk of lung cancer for patients who lived in areas with high nitrogen oxide concentrations. In this study, the association was higher for non-smokers than smokers. An additional Danish study, also in 2011, likewise noted evidence of possible associations between air pollution and other forms of cancer, including cervical cancer and brain cancer.

In December 2015, medical scientists reported that cancer is overwhelmingly a result of environmental factors, and not largely down to bad luck. Maintaining a healthy weight, eating a healthy diet, minimizing alcohol and eliminating smoking reduces the risk of developing the disease, according to the researchers.

Children

In the United States, despite the passage of the Clean Air Act in 1970, in 2002 at least 146 million Americans were living in non-attainment areas—regions in which the concentration of certain air pollutants exceeded federal standards. These dangerous pollutants are known as the criteria pollutants, and include ozone, particulate matter, sulphur dioxide, nitrogen dioxide, carbon monoxide, and lead. Protective measures to ensure children's health are being taken in cities such as New Delhi, India where buses now use compressed natural gas to help eliminate the "pea-soup" smog. A recent study in Europe has found that exposure to ultrafine particles can increase blood pressure in children. According to a WHO report-2018, polluted air is a main cause poisoning millions of children under the age of 15 years and ruining their lives which resulting to death of some six hundred thousand children annually.

Infants

Ambient levels of air pollution have been associated with preterm birth and low birth weight. A 2014 WHO worldwide survey on maternal and perinatal health found a statistically significant association between low birth weights (LBW) and increased levels of exposure to PM2.5. Women in regions with greater than average PM2.5 levels had statistically significant higher odds of pregnancy resulting in a low-birth weight infant even when adjusted for country-related variables. The effect is thought to be from stimulating inflammation and increasing oxidative stress.

A study by the University of York found that in 2010 exposure to PM2.5 was strongly associated with 18% of preterm births globally, which was approximately 2.7 million premature births. The countries with the highest air pollution associated preterm births were in South and East Asia, the Middle East, North Africa, and West sub-Saharan Africa.

The source of PM 2.5 differs greatly by region. In South and East Asia, pregnant women are frequently exposed to indoor air pollution because of the wood and other biomass fuels used for cooking which are responsible for more than 80% of regional pollution. In the Middle East, North Africa and West sub-Saharan Africa, fine PM comes from natural sources, such as dust storms. The United States had an estimated 50,000 preterm births associated with exposure to PM2.5 in 2010.

A study performed by Wang, et al. between the years of 1988 and 1991 has found a correlation between sulphur Dioxide (SO_2) and total suspended particulates (TSP) and preterm births and low birth weights in Beijing. A group of 74,671 pregnant women, in four separate regions of Beijing, were monitored from early pregnancy to delivery along with daily air pollution levels of sulphur Dioxide and TSP (along with other particulates). The estimated reduction in birth weight was 7.3 g for every 100 $\mu g/m^3$ increase in SO_2 and 6.9g for each 100 $\mu g/m^3$ increase in TSP. These associations were statistically significant in both summer and winter, although, summer was greater. The proportion of low birth weight attributable to air pollution, was 13%. This is the largest attributable risk ever reported for the known risk factors of low birth weight. Coal stoves, which are in 97% of homes, are a major source of air pollution in this area.

Brauer et al. studied the relationship between air pollution and proximity to a highway with pregnancy outcomes in a Vancouver cohort of pregnant woman using addresses to estimate exposure during pregnancy. Exposure to NO, NO_2, CO PM10 and PM2.5 were associated with infants born small for gestational age (SGA). Women living <50meters away from an expressway or highway were 26% more likely to give birth to a SGA infant.

Clean Areas

Even in the areas with relatively low levels of air pollution, public health effects can be significant and costly, since a large number of people breathe in such pollutants. A study published in 2017 found that even in areas of the U.S. where ozone and PM2.5 meet federal standards, Medicare recipients who are exposed to more air pollution have higher mortality rates. A 2005 scientific study for the British Columbia Lung Association showed that a small improvement in air quality (1% reduction of ambient PM2.5 and ozone concentrations) would produce $29 million in annual savings in the Metro Vancouver region in 2010. This finding is based on health valuation of lethal (death) and sub-lethal (illness) affects.

Central Nervous System

Data is accumulating that air pollution exposure also affects the central nervous system.

In a June 2014 study conducted by researchers, it was discovered that early exposure to air pollution causes the same damaging changes in the brain as autism and schizophrenia. The study also shows that air pollution also affected short-term memory, learning ability, and impulsivity. Lead researcher Professor Deborah Cory-Slechta said that "When we looked closely at the ventricles, we could see that the white matter that normally surrounds them hadn't fully developed. It appears that inflammation had damaged those brain cells and prevented that region of the brain from developing, and the ventricles simply expanded to fill the space. Our findings add to the growing body of evidence that air pollution may play a role in autism, as well as in other neurodevelopmental disorders." Air pollution has a more significant negative effect on males than on females.

In 2015, experimental studies reported the detection of significant episodic (situational) cognitive impairment from impurities in indoor air breathed by test subjects who were not informed about changes in the air quality. Researchers at the Harvard University and SUNY Upstate Medical University and Syracuse University measured the cognitive performance of 24 participants in three different controlled laboratory atmospheres that simulated those found in "conventional" and "green" buildings, as well as green buildings with enhanced ventilation. Performance was evaluated objectively using the widely used Strategic Management Simulation software simulation tool, which is a well-validated assessment test for executive decision-making in an unconstrained situation allowing initiative and improvisation. Significant deficits were observed in the performance scores achieved in increasing concentrations of either volatile organic compounds (VOCs) or carbon dioxide, while keeping other factors constant. The highest impurity levels reached are not uncommon in some classroom or office environments. Air pollution increases the risk of dementia in people over 50 years old.

Reduction Efforts

Various air pollution control technologies and strategies are available to reduce air pollution. At its most basic level, land-use planning is likely to involve zoning and transport infrastructure planning. In most developed countries, land-use planning is an important part of social policy, ensuring that land is used efficiently for the benefit of the wider economy and population, as well as to protect the environment.

Because a large share of air pollution is caused by combustion of fossil fuels such as coal and oil, the reduction of these fuels can reduce air pollution drastically. Most effective is the switch to clean power sources such as wind power, solar power, hydro power which don't cause air pollution. Efforts to reduce pollution from mobile sources includes primary regulation (many developing countries have permissive regulations), expanding regulation to new sources (such as cruise and transport ships, farm equipment, and small gas-powered equipment such as string trimmers, chainsaws, and snowmobiles), increased fuel efficiency (such as through the use of hybrid vehicles), conversion to cleaner fuels or conversion to electric vehicles.

Titanium dioxide has been researched for its ability to reduce air pollution. Ultraviolet light will release free electrons from material, thereby creating free radicals, which break up VOCs and NOx gases. One form is superhydrophilic.

In 2014, Prof. Tony Ryan and Prof. Simon Armitage of University of Sheffield prepared a 10 meter by 20 meter-sized poster coated with microscopic, pollution-eating nanoparticles of titanium dioxide. Placed on a building, this giant poster can absorb the toxic emission from around 20 cars each day.

A very effective means to reduce air pollution is the transition to renewable energy. According to a study published in Energy and Environmental Science in 2015 the switch to 100% renewable energy in the United States would eliminate about 62,000 premature mortalities per year and about 42,000 in 2050, if no biomass were used. This would save about $600 billion in health costs a year due to reduced air pollution in 2050, or about 3.6% of the 2014 U.S. gross domestic product.

Control Devices

The following items are commonly used as pollution control devices in industry and transportation. They can either destroy contaminants or remove them from an exhaust stream before it is emitted into the atmosphere.

- Particulate control:
    - Mechanical collectors (dust cyclones, multicyclones).
    - Electrostatic precipitators An electrostatic precipitator (ESP), or electrostatic air cleaner is a particulate collection device that removes particles

from a flowing gas (such as air), using the force of an induced electrostatic charge. Electrostatic precipitators are highly efficient filtration devices that minimally impede the flow of gases through the device, and can easily remove fine particulates such as dust and smoke from the air stream.

- Baghouses Designed to handle heavy dust loads, a dust collector consists of a blower, dust filter, a filter-cleaning system, and a dust receptacle or dust removal system (distinguished from air cleaners which utilize disposable filters to remove the dust).
- Particulate scrubbers Wet scrubber is a form of pollution control technology. The term describes a variety of devices that use pollutants from a furnace flue gas or from other gas streams. In a wet scrubber, the polluted gas stream is brought into contact with the scrubbing liquid, by spraying it with the liquid, by forcing it through a pool of liquid, or by some other contact method, so as to remove the pollutants.

- Scrubbers:
  - Baffle spray scrubber.
  - Cyclonic spray scrubber.
  - Ejector venturi scrubber.
  - Mechanically aided scrubber.
  - Spray tower.
  - Wet scrubber.
- NO_x control:
  - Low NO_x burners.
  - Selective catalytic reduction (SCR).
  - Selective non-catalytic reduction (SNCR).
  - NO_x scrubbers.
  - Exhaust gas recirculation.
  - Catalytic converter (also for VOC control).
- VOC abatement:
  - Adsorption systems, using activated carbon, such as Fluidized Bed Concentrator.

- Flares.
- Thermal oxidizers.
- Catalytic converters.
- Biofilters.
- Absorption (scrubbing).
- Cryogenic condensers.
- Vapor recovery systems.

- Acid Gas/SO_2 control:
    - Wet scrubbers.
    - Dry scrubbers.
    - Flue-gas desulphurization.
- Mercury control:
    - Sorbent Injection Technology.
    - Electro-Catalytic Oxidation (ECO).
    - K-Fuel.
- Dioxin and furan control.
- Miscellaneous associated equipment:
    - Source capturing systems.
    - Continuous emissions monitoring systems (CEMS).

Hotspots

Air pollution hotspots are areas where air pollution emissions expose individuals to increased negative health effects. They are particularly common in highly populated, urban areas, where there may be a combination of stationary sources (e.g. industrial facilities) and mobile sources (e.g. cars and trucks) of pollution. Emissions from these sources can cause respiratory disease, childhood asthma, cancer, and other health problems. Fine particulate matter such as diesel soot, which contributes to more than 3.2 million premature deaths around the world each year, is a significant problem. It is very small and can lodge itself within the lungs and enter the bloodstream. Diesel soot is concentrated in densely populated areas, and one in six people in the U.S. live near a diesel pollution hot spot.

While air pollution hotspots affect a variety of populations, some groups are more likely to be located in hotspots. Previous studies have shown disparities in exposure to pollution by race and/or income. Hazardous land uses (toxic storage and disposal facilities, manufacturing facilities, major roadways) tend to be located where property values and income levels are low. Low socioeconomic status can be a proxy for other kinds of social vulnerability, including race, a lack of ability to influence regulation and a lack of ability to move to neighborhoods with less environmental pollution. These communities bear a disproportionate burden of environmental pollution and are more likely to face health risks such as cancer or asthma.

Studies show that patterns in race and income disparities not only indicate a higher exposure to pollution but also higher risk of adverse health outcomes. Communities characterized by low socioeconomic status and racial minorities can be more vulnerable to cumulative adverse health impacts resulting from elevated exposure to pollutants than more privileged communities. Blacks and Latinos generally face more pollution than whites and Asians, and low-income communities bear a higher burden of risk than affluent ones. Racial discrepancies are particularly distinct in suburban areas of the Southern United States and metropolitan areas of the Midwestern and Western United States. Residents in public housing, who are generally low-income and cannot move to healthier neighborhoods, are highly affected by nearby refineries and chemical plants.

Particulates

Particulates – also known as atmospheric aerosol particles, atmospheric particulate matter, particulate matter (PM), or suspended particulate matter (SPM) – are microscopic solid or liquid matter suspended in the air. The term *aerosol* commonly refers to the particulate/air mixture, as opposed to the particulate matter alone. Sources of particulate matter can be natural or anthropogenic. They have impacts on climate and precipitation that adversely affect human health, in addition to direct inhalation.

Subtypes of atmospheric particles include suspended particulate matter (SPM), thoracic and respirable particles, inhalable coarse particles, which are coarse particles with a diameter between 2.5 and 10 micrometers (μm) (PM_{10}), fine particles with a diameter of 2.5 μm or less ($PM_{2.5}$), ultrafine particles, and soot.

The IARC and WHO designate airborne particulates a Group 1 carcinogen. Particulates are the most harmful form of air pollution due to their ability to penetrate deep into the lungs and blood streams unfiltered, causing heart attacks, respiratory disease, and premature death. In 2013, a study involving 312,944 people in nine European countries revealed that there was no safe level of particulates and that for every increase of 10 $\mu g/m^3$ in PM_{10}, the lung cancer rate rose 22%. The smaller $PM_{2.5}$ were particularly deadly, with a 36% increase in lung cancer per 10 $\mu g/m^3$ as it can penetrate deeper into the lungs. Worldwide exposure to $PM_{2.5}$ contributed to 4.1 million deaths from heart disease and stroke, lung cancer, chronic lung disease, and respiratory infections in 2016.

Overall, ambient particulate matter ranks as the sixth leading risk factor for premature death globally.

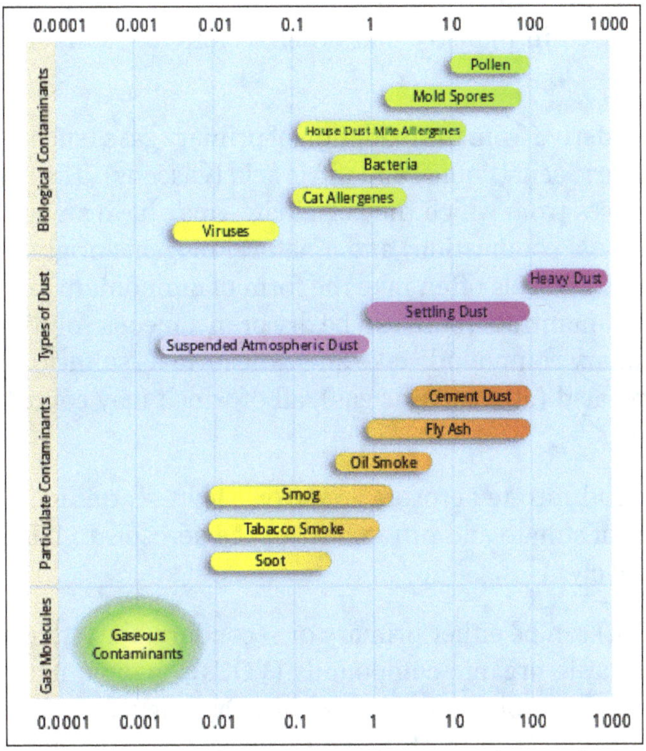

This diagram shows types, and size distribution in micrometres, of atmospheric particulate matter.

Sources of Atmospheric Particulate Matter

Some particulates occur naturally, originating from volcanoes, dust storms, forest and grassland fires, living vegetation and sea spray. Human activities, such as the burning of fossil fuels in vehicles, stubble burning, power plants, road dust, wet cooling towers in cooling systems and various industrial processes, also generate significant amounts of particulates. Coal combustion in developing countries is the primary method for heating homes and supplying energy. Because salt spray over the oceans is the overwhelmingly most common form of particulate in the atmosphere, *anthropogenic* aerosols—those made by human activities—currently account for about 10 percent of the total mass of aerosols in our atmosphere.

Composition

The composition of aerosols and particles depends on their source. Wind-blown mineral dust tends to be made of mineral oxides and other material blown from the Earth's crust; this particulate is light-absorbing. Sea salt is considered the second-largest contributor in the global aerosol budget, and consists mainly of sodium chloride originated from sea spray; other constituents of atmospheric sea salt reflect the composition of sea water, and thus include magnesium, sulfate, calcium, potassium, etc. In addition, sea spray aerosols may contain organic compounds, which influence their chemistry. The drift/mist emissions from the wet cooling towers is also source of particulate matter as they are widely used in industry and other sectors for dissipating heat in cooling systems.

Secondary particles derive from the oxidation of primary gases such as sulfur and nitrogen oxides into sulfuric acid (liquid) and nitric acid (gaseous). The precursors for these aerosols—i.e. the gases from which they originate—may have an anthropogenic origin (from fossil fuel or coal combustion) and a natural biogenic origin. In the presence of ammonia, secondary aerosols often take the form of ammonium salts; i.e. ammonium sulfate and ammonium nitrate (both can be dry or in aqueous solution); in the absence of ammonia, secondary compounds take an acidic form as sulfuric acid (liquid aerosol droplets) and nitric acid (atmospheric gas), all of which may contribute to the health effects of particulates.

Secondary sulfate and nitrate aerosols are strong light-scatterers. This is mainly because the presence of sulfate and nitrate causes the aerosols to increase to a size that scatters light effectively.

Organic matter (OM) can be either primary or secondary, the latter part deriving from the oxidation of volatile organic compounds (VOCs); organic material in the atmosphere may either be biogenic or anthropogenic. Organic matter influences the atmospheric radiation field by both scattering and absorption. Another important aerosol type is elemental carbon (EC, also known as *black carbon*, BC): this aerosol type includes strongly light-absorbing material and is thought to yield large positive radiative forcing. Organic matter and elemental carbon together constitute the carbonaceous fraction of aerosols. Secondary organic aerosols (SOAs), tiny "tar balls" resulting from combustion products of internal combustion engines, have been identified as a danger to health.

The chemical composition of the aerosol directly affects how it interacts with solar radiation. The chemical constituents within the aerosol change the overall refractive index. The refractive index will determine how much light is scattered and absorbed.

The composition of particulate matter that generally causes visual effects such as smog consists of sulfur dioxide, nitrogen oxides, carbon monoxide, mineral dust, organic matter, and elemental carbon also known as black carbon or soot. The particles are hygroscopic due to the presence of sulfur, and SO_2 is converted to sulfate when high humidity and low temperatures are present. This causes the reduced visibility and yellow color.

Size Distribution of Particulates

Aerosol particles of natural origin (such as windblown dust) tend to have a larger radius than human-produced aerosols such as particle pollution. The false-color maps in human pollution, or a mixture of both, monthly.

Among the most obvious patterns that the size distribution time series shows is that in the planet's most southerly latitudes, nearly all the aerosols are large, but in the high northern latitudes, smaller aerosols are very abundant. Most of the Southern Hemisphere is covered by ocean, where the largest source of aerosols is natural sea salt from dried sea spray. Because the land is concentrated in the Northern Hemisphere, the amount of small aerosols from fires and human activities is greater there than in the Southern Hemisphere. Over land, patches of large-radius aerosols appear over deserts and arid regions, most prominently, the Sahara Desert in North Africa and the Arabian Peninsula, where dust storms are common. Places where human-triggered or natural fire activity is common (land-clearing fires in the Amazon from August–October, for example, or lightning-triggered fires in the forests of northern Canada in Northern Hemisphere summer) are dominated by smaller aerosols. Human-produced (fossil fuel) pollution is largely responsible for the areas of small aerosols over developed areas such as the eastern United States and Europe, especially in their summer.

Satellite measurements of aerosols, called aerosol optical thickness, are based on the fact that the particles change the way the atmosphere reflects and absorbs visible and infrared light. an optical thickness of less than 0.1 (palest yellow) indicates a crystal clear sky with maximum visibility, whereas a value of 1 (reddish brown) indicates very hazy conditions.

Deposition Processes

In general, the smaller and lighter a particle is, the longer it will stay in the air. Larger particles (greater than 10 micrometers in diameter) tend to settle to the ground by gravity in a matter of hours whereas the smallest particles (less than 1 micrometer) can stay in the atmosphere for weeks and are mostly removed by precipitation. Diesel particulate matter is highest near the source of emission. Any info regarding DPM and the atmosphere, flora, height, and distance from major sources would be useful to determine health effects.

Controlling Technologies

A complicated blend of solid and liquid particles result in particulate matter and these particulate matter emissions are highly regulated in most industrialized countries. Due to environmental concerns, most industries are required to operate some kind of dust collection system to control particulate emissions. These systems include inertial collectors (cyclonic separators), fabric filter collectors (baghouses), wet scrubbers, and electrostatic precipitators.

Cyclonic separators are useful for removing large, coarse particles and are often employed as a first step or "pre-cleaner" to other more efficient collectors. Well-designed cyclonic separators can be very efficient in removing even fine particulates, and may be operated continuously without requiring frequent shutdowns for maintenance.

Fabric filters or baghouses are the most commonly employed in general industry. They work by forcing dust laden air through a bag shaped fabric filter leaving the particulate to collect on the outer surface of the bag and allowing the now clean air to pass through to either be exhausted into the atmosphere or in some cases recirculated into the facility. Common fabrics include polyester and fiberglass and common fabric coatings include PTFE (commonly known as Teflon). The excess dust buildup is then cleaned from the bags and removed from the collector.

Wet scrubbers pass the dirty air through a scrubbing solution (usually a mixture of water and other compounds) allowing the particulate to attach to the liquid molecules. Electrostatic precipitators electrically charge the dirty air as it passes through. The now charged air then passes through large electrostatic plates which attract the charged particle in the airstream collecting them and leaving the now clean air to be exhausted or recirculated.

Besides removing particulates from the source of the pollution, it can also be cleaned in the open air.

Climate Effects

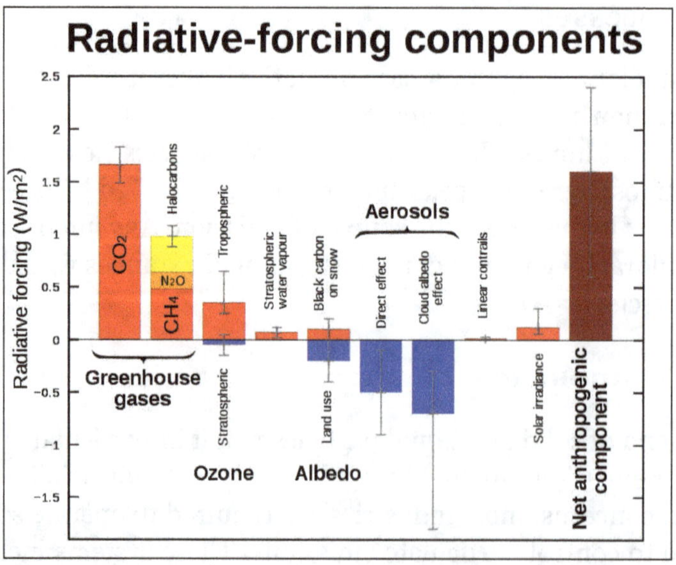

2005 radiative forcings and uncertainties as estimated by the IPCC.

Atmospheric aerosols affect the climate of the earth by changing the amount of incoming solar radiation and outgoing terrestrial longwave radiation retained in the earth's system. This occurs through several distinct mechanisms which are split into direct, indirect and semi-direct aerosol effects. The aerosol climate effects are the biggest source of uncertainty in future climate predictions. The Intergovernmental Panel on Climate Change, Third Assessment Report, says: While the radiative forcing due to greenhouse gases may be determined to a reasonably high degree of accuracy the uncertainties relating to aerosol radiative forcings remain large, and rely to a large extent on the estimates from global modelling studies that are difficult to verify at the present time.

Aerosol Radiative Effects

Direct Effect

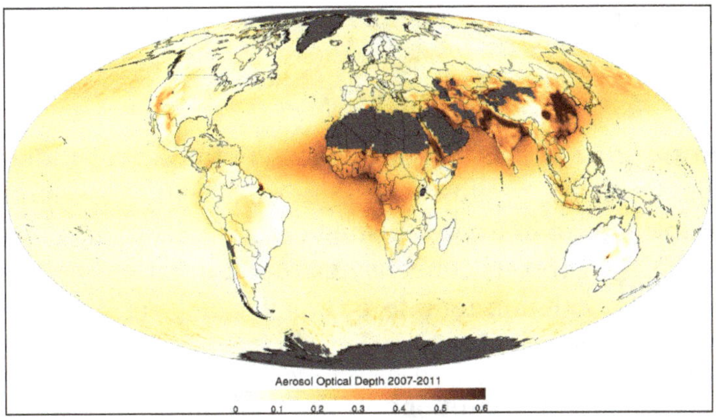

Global aerosol optical thickness. The aerosol scale (yellow to dark reddish-brown) indicates the relative amount of particles that absorb sunlight.

The direct aerosol effect consists of any direct interaction of radiation with atmospheric aerosols, such as absorption or scattering. It affects both short and longwave radiation to produce a net negative radiative forcing. The magnitude of the resultant radiative forcing due to the direct effect of an aerosol is dependent on the albedo of the underlying surface, as this affects the net amount of radiation absorbed or scattered to space. e.g. if a highly scattering aerosol is above a surface of low albedo it has a greater radiative forcing than if it was above a surface of high albedo. The converse is true of absorbing aerosol, with the greatest radiative forcing arising from a highly absorbing aerosol over a surface of high albedo. The direct aerosol effect is a first order effect and is therefore classified as a radiative forcing by the IPCC. The interaction of an aerosol with radiation is quantified by the single-scattering albedo (SSA), the ratio of scattering alone to scattering plus absorption (*extinction*) of radiation by a particle. The SSA tends to unity if scattering dominates, with relatively little absorption, and decreases as absorption increases, becoming zero for infinite absorption. For example, the sea-salt aerosol has an SSA of 1, as a sea-salt particle only scatters, whereas soot has an SSA of 0.23, showing that it is a major atmospheric aerosol absorber.

Particulates in the air causing shades of grey and pink in Mumbai during sunset.

Indirect Effect

The Indirect aerosol effect consists of any change to the earth's radiative budget due to the modification of clouds by atmospheric aerosols, and consists of several distinct effects. Cloud droplets form onto pre-existing aerosol particles, known as cloud condensation nuclei (CCN). Droplets condensing around human-produced aerosols such as found in particulate pollution tend to be smaller and more numerous than those forming around aerosol particles of natural origin (such as windblown dust).

For any given meteorological conditions, an increase in CCN leads to an increase in the number of cloud droplets. This leads to more scattering of shortwave radiation i.e. an increase in the albedo of the cloud, known as the Cloud albedo effect, First indirect effect or Twomey effect. Evidence supporting the cloud albedo effect has been observed from the effects of ship exhaust plumes and biomass burning on cloud albedo compared to ambient clouds. The Cloud albedo aerosol effect is a first order effect and therefore classified as a radiative forcing by the IPCC.

An increase in cloud droplet number due to the introduction of aerosol acts to reduce the cloud droplet size, as the same amount of water is divided into more droplets. This has the effect of suppressing precipitation, increasing the cloud lifetime, known as the cloud lifetime aerosol effect, second indirect effect or Albrecht effect. This has been observed as the suppression of drizzle in ship exhaust plume compared to ambient clouds, and inhibited precipitation in biomass burning plumes. This cloud lifetime effect is classified as a climate feedback (rather than a radiative forcing) by the IPCC due to the interdependence between it and the hydrological cycle. However, it has previously been classified as a negative radiative forcing.

Semi-direct Effect

The Semi-direct effect concerns any radiative effect caused by absorbing atmospheric aerosol such as soot, apart from direct scattering and absorption, which is classified as the direct effect. It encompasses many individual mechanisms, and in general is more poorly defined and understood than the direct and indirect aerosol effects. For instance, if absorbing aerosols are present in a layer aloft in the atmosphere, they can heat surrounding air which inhibits the condensation of water vapour, resulting in less cloud formation. Additionally, heating a layer of the atmosphere relative to the surface results in a more stable atmosphere due to the inhibition of atmospheric convection. This inhibits the convective uplift of moisture, which in turn reduces cloud formation. The heating of the atmosphere aloft also leads to a cooling of the surface, resulting in less evaporation of surface water. The effects described here all lead to a reduction in cloud cover i.e. an increase in planetary albedo. The semi-direct effect classified as a climate feedback) by the IPCC due to the interdependence between it and the hydrological cycle. However, it has previously been classified as a negative radiative forcing.

Roles of Different Aerosol Species

Sulfate Aerosol

Sulfate aerosol has two main effects, direct and indirect. The direct effect, via albedo, is a cooling effect that slows the overall rate of global warming: the IPCC's best estimate of the radiative forcing is −0.4 watts per square meter with a range of −0.2 to −0.8 W/m^2 but there are substantial uncertainties. The effect varies strongly geographically, with most cooling believed to be at and downwind of major industrial centres. Modern climate models addressing the attribution of recent climate change take into account sulfate forcing, which appears to account (at least partly) for the slight drop in global temperature in the middle of the 20th century. The indirect effect (via the aerosol acting as cloud condensation nuclei, CCN, and thereby modifying the cloud properties -albedo and lifetime-) is more uncertain but is believed to be a cooling.

Black Carbon

Black carbon (BC), or carbon black, or elemental carbon (EC), often called soot, is composed of pure carbon clusters, skeleton balls and buckyballs, and is one of the most important absorbing aerosol species in the atmosphere. It should be distinguished from organic carbon (OC): clustered or aggregated organic molecules on their own or permeating an EC buckyball. BC from fossil fuels is estimated by the IPCC in the Fourth Assessment Report of the IPCC, 4AR, to contribute a global mean radiative forcing of +0.2 W/m^2 (was +0.1 W/m^2 in the Second Assessment Report of the IPCC, SAR), with a range +0.1 to +0.4 W/m^2. Bond et al., however, states that "the best estimate for the industrial-era (1750 to 2005) direct radiative forcing of atmospheric black carbon is +0.71 W/m^2 with 90% uncertainty bounds of (+0.08, +1.27) W/m^2" with "total direct forcing by all black carbon sources, without subtracting the preindustrial background, is estimated as +0.88 (+0.17, +1.48) W/m^2".

Instances of Aerosol affecting Climate

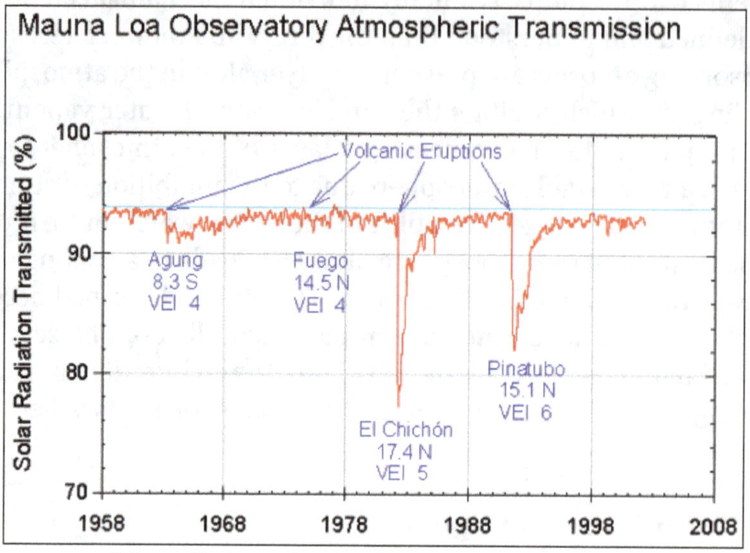

Solar radiation reduction due to volcanic eruptions.

Volcanoes are a large natural source of aerosol and have been linked to changes in the earth's climate often with consequences for the human population. Eruptions linked to changes in climate include the 1600 eruption of Huaynaputina which was linked to the Russian famine of 1601 - 1603, leading to the deaths of two million, and the 1991 eruption of Mount Pinatubo which caused a global cooling of approximately 0.5 °C lasting several years. Research tracking the effect of light-scattering aerosols in the stratosphere during 2000 and 2010 and comparing its pattern to volcanic activity show a close correlation. Simulations of the effect of anthropogenic particles showed little influence at present levels.

Aerosols are also thought to affect weather and climate on a regional scale. The failure of the Indian Monsoon has been linked to the suppression of evaporation of water from the Indian Ocean due to the semi-direct effect of anthropogenic aerosol.

Recent studies of the Sahel drought and major increases since 1967 in rainfall over the Northern Territory, Kimberley, Pilbara and around the Nullarbor Plain have led some scientists to conclude that the aerosol haze over South and East Asia has been steadily shifting tropical rainfall in both hemispheres southward.

The latest studies of severe rainfall decline over southern Australia since 1997 have led climatologists there to consider the possibility that these Asian aerosols have shifted not only tropical but also midlatitude systems southward.

Health Effects

Size, Shape and Solubility Matter

The size of the particle is a main determinant of where in the respiratory tract the particle will come to rest when inhaled. Larger particles are generally filtered in the nose and throat via cilia and mucus, but particulate matter smaller than about 10 micrometers, can settle in the bronchi and lungs and cause health problems. The 10-micrometer size does not represent a strict boundary between respirable and non-respirable particles, but has been agreed upon for monitoring of airborne particulate matter by most regulatory agencies. Because of their small size, particles on the order of 10 micrometers or less (coarse particulate matter, PM_{10}) can penetrate the deepest part of the lungs such as the bronchioles or alveoli; when asthmatics are exposed to these conditions it can trigger bronchoconstriction.

Air pollution measurement station in Emden, Germany.

Similarly, so called fine particulate matter ($PM_{2.5}$), tend to penetrate into the gas exchange regions of the lung (alveolus), and very small particles (ultrafine particulate matter, $PM_{0.1}$) may pass through the lungs to affect other organs. Penetration of particles is not wholly dependent on their size; shape and chemical composition also play a part. To avoid this complication, simple nomenclature is used to indicate the different degrees of relative penetration of a PM particle into the cardiovascular system. Inhalable particles penetrate no further than the bronchi as they are filtered out by the cilia. Thoracic particles can penetrate right into terminal bronchioles whereas $PM_{0.1}$, which can penetrate to alveoli, the gas exchange area, and hence the circulatory system are termed respirable particles. In analogy, the inhalable dust fraction is the fraction of dust entering nose and mouth which may be deposited anywhere in the respiratory tract. The thoracic fraction is the fraction that enters the thorax and is deposited within the lung's airways. The respirable fraction is what is deposited in the gas exchange regions (alveoli).

The smallest particles, less than 100 nanometers (nanoparticles), may be even more damaging to the cardiovascular system. Nanoparticles can pass through cell membranes and migrate into other organs, including the brain. Particles emitted from modern diesel engines (commonly referred to as Diesel Particulate Matter, or DPM) are typically in the size range of 100 nanometers (0.1 micrometer). These soot particles also carry carcinogens like benzopyrenes adsorbed on their surface. Particulate mass is not a proper measure of the health hazard, because one particle of 10 μm diameter has approximately the same mass as 1 million particles of 100 nm diameter, but is much less hazardous, as it is unlikely to enter the alveoli. Legislative limits for engine emissions based on mass are therefore not protective. Proposals for new regulations exist in some countries,with suggestions to limit the particle *surface area* or the *particle count* (numerical quantity) instead.

The site and extent of absorption of inhaled gases and vapors are determined by their solubility in water. Absorption is also dependent upon air flow rates and the partial pressure of the gases in the inspired air. The fate of a specific contaminant is dependent upon the form in which it exists (aerosol or particulate). Inhalation also depends upon the breathing rate of the subject.

Another complexity not entirely documented is how the shape of PM can affect health, except for the needle-like shape of asbestos which can lodge itself in the lungs. Geometrically angular shapes have more surface area than rounder shapes, which in turn affects the binding capacity of the particle to other, possibly more dangerous substances.

Health Problems

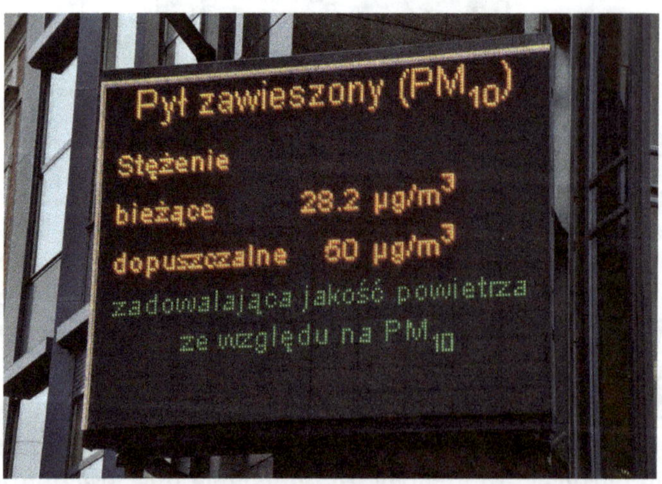

Air quality information on PM10 displayed in Katowice, Poland.

The effects of inhaling particulate matter that has been widely studied in humans and animals include asthma, lung cancer, respiratory diseases, cardiovascular disease, premature delivery, birth defects, low birth weight, and premature death.

Inhalation of $PM_{2.5} - PM_{10}$ is associated with elevated risk of adverse pregnancy outcomes, such as low birth weight . Maternal $PM_{2.5}$ exposure during pregnancy is also associated with high blood pressure in children. Exposure to $PM_{2.5}$ has been associated with greater reductions in birth weight than exposure to PM_{10}. PM exposure can cause inflammation, oxidative stress, endocrine disruption, and impaired oxygen transport access to the placenta, all of which are mechanisms for heightening the risk of low birth weight. Overall epidemiologic and toxicological evidence suggests that a causal relationship exists between long-term exposures to $PM_{2.5}$ and developmental outcomes (i.e. low birth weight). However, studies investigating the significance of trimester-specific exposure have proven to be inconclusive, and results of international studies have been inconsistent in drawing associations of prenatal particulate matter exposure and low birth weight. As perinatal outcomes have been associated with lifelong health and exposure to particulate matter is widespread, this issue is of critical public health importance and additional research will be essential to inform public policy on the matter.

Increased levels of fine particles in the air as a result of *anthropogenic* particulate air pollution "is consistently and independently related to the most serious effects, including lung cancer and other cardiopulmonary mortality." A large number of deaths and other health problems associated with particulate pollution was first demonstrated in the early 1970s and has been reproduced many times since. PM pollution is estimated to cause 22,000–52,000 deaths per year in the United States (from 2000) contributed to ~370,000 premature deaths in Europe during 2005. and 3.22 million deaths globally in 2010 per the global burden of disease collaboration.

A 2002 study indicated that $PM_{2.5}$ leads to high plaque deposits in arteries, causing vascular inflammation and atherosclerosis – a hardening of the arteries that reduces elasticity, which can lead to heart attacks and other cardiovascular problems. A 2014 meta analysis reported that long term exposure to particulate matter is linked to coronary events. The study included 11 cohorts participating in the European Study of Cohorts for Air Pollution Effects (ESCAPE) with 100,166 participants, followed for an average of 11.5 years. An increase in estimated annual exposure to PM 2.5 of just 5 $\mu g/m^3$ was linked with a 13% increased risk of heart attacks. In 2017 a study revealed that PM not only affects human cells and tissues, but also impacts bacteria which cause disease in humans. This study concluded that biofilm formation, antibiotic tolerance, and colonisation of both *Staphylococcus aureus* and *Streptococcus pneumoniae* was altered by Black Carbon exposure.

The World Health Organization (WHO) estimated in 2005 that "fine particulate air pollution (PM(2.5)), causes about 3% of mortality from cardiopulmonary disease, about 5% of mortality from cancer of the trachea, bronchus, and lung, and about 1% of mortality from acute respiratory infections in children under 5 years, worldwide". A 2011 study concluded that traffic exhaust is the single most serious preventable cause of heart attack in the general public, the cause of 7.4% of all attacks.

The largest US study on acute health effects of coarse particle pollution between 2.5 and

10 micrometers in diameter. was published 2008 and found an association with hospital admissions for cardiovascular diseases but no evidence of an association with the number of hospital admissions for respiratory diseases. After taking into account fine particle levels ($PM_{2.5}$ and less), the association with coarse particles remained but was no longer statistically significant, which means the effect is due to the subsection of fine particles.

Particulate matter studies in Bangkok Thailand from 2008 indicated a 1.9% increased risk of dying from cardiovascular disease, and 1.0% risk of all disease for every 10 micrograms per cubic meter. Levels averaged 65 in 1996, 68 in 2002, and 52 in 2004. Decreasing levels may be attributed to conversions of diesel to natural gas combustion as well as improved regulations.

The Mongolian government agency recorded a 45% increase in the rate of respiratory illness in the past five years (reported in September 2014). Bronchial asthma, chronic obstructive pulmonary disease and interstitial pneumonia were the most common ailments treated by area hospitals. Levels of premature death, chronic bronchitis, and cardiovascular disease are increasing at a rapid rate.

A study In 2000 conducted in the U.S. explored how fine particulate matter may be more harmful than coarse particulate matter. The study was based on six different cities. They found that deaths and hospital visits that were caused by particulate matter in the air were primarily due fine particulate matter.

Effects on Vegetation

Particulate matter can clog stomatal openings of plants and interfere with photosynthesis functions. In this manner, high particulate matter concentrations in the atmosphere can lead to growth stunting or mortality in some plant species.

AIR POLLUTION FORECASTING

Air pollution forecasting is the application of science and technology to predict the composition of the Air pollution in the atmosphere for a given location and time.

The forecast may give the pollutants concentration or the air quality index.

Countries and cities are given forecasts by state and local government organizations, as well as private companies like Airly, AirVisual, Aerostate, BreezoMeter, PlumeLabs, and DRAXIS that give air pollution forecast.

Techniques

- Air pollution forecasting can be done by coupling weather forecasting systems with Chemical transport model and Atmospheric dispersion modeling.

- Using machine learning techniques.

The forecast takes into account local emission sources (like nearby traffic or industry) and remote sources (e.g. dust that is carried by air parcels and follows the wind direction).

The forecast temporally resolution is usually daily or hourly and the spatial resolution can change from block resolution to dozens of km resolution.

Most forecasts of air quality cover two to five days.

AIR QUALITY MONITORING

Air Quality Monitoring is the process of examining the quality of air in the atmosphere. The starting point of air quality monitoring is to first study if an area has an air pollution problem. Monitoring helps in assessing the level of pollution in relation to the ambient air quality standards. Standards are a regulatory measure to set the target for pollution reduction and achieve clean air. Robust monitoring helps to guard against extreme events by alerting people and initiate action. We regulate a total of 12 pollutants, including SO_2, NO_2, PM_{10}, $PM_{2.5}$ (particulate matter of up to 10 micron and up to 2.5 micron size), ozone, lead, arsenic, nickel, CO, NH_3, benzene, and BaP (particulate phase). Across cities, only SO_2, NO_2 and RSPM/PM10 are monitored regularly. Other pollutants, such as $PM_{2.5}$, O_3, CO, BTX, heavy metals are monitored in select cities as capacity is still being built.

Ventilation (Architecture)

An ab anbar (water reservoir) with double domes and windcatchers (openings near the top of the towers) in the central desert city of Naeen, Iran. Windcatchers are a form of natural ventilation.

Ventilation is the intentional introduction of outdoor air into a space and is mainly used to control indoor air quality by diluting and displacing indoor pollutants; it can also be used for purposes of thermal comfort or dehumidification.

The intentional introduction of outdoor air can be categorized as either mechanical ventilation, or natural ventilation. Mechanical ventilation uses fans to drive the flow of outdoor air into a building. This may be accomplished by pressurization (in the case of positively pressurized buildings), or by depressurization (in the case of exhaust ventilation systems). Many mechanically ventilated buildings use a combination of both, with the ventilation being integrated into the HVAC system. Natural ventilation is the intentional passive flow of outdoor air into a building through planned openings (such as louvers, doors, and windows). Natural ventilation does not require mechanical systems to move outdoor air, it relies entirely on passive physical phenomena, such as diffusion, wind pressure, or the stack effect. Mixed mode ventilation systems use both mechanical and natural processes. The mechanical and natural components may be used in conjunction with each other or separately at different times of day or season of the year. Since the natural component can be affected by unpredictable environmental conditions it may not always provide an appropriate amount of ventilation. In this case, mechanical systems may be used to supplement or to regulate the naturally driven flow.

In many instances, ventilation for indoor air quality is simultaneously beneficial for the control of thermal comfort. At these times, it can be useful to increase the rate of ventilation beyond the minimum required for indoor air quality. Two examples include air-side economizer strategies and ventilative pre-cooling. In other instances, ventilation for indoor air quality contributes to the need for - and energy use by - mechanical heating and cooling equipment. In hot and humid climates, dehumidification of ventilation air can be a particularly energy intensive process.

Ventilation should be considered for its relationship to "venting" for appliances and combustion equipment such as water heaters, furnaces, boilers, and wood stoves. Most importantly, the design of building ventilation must be careful to avoid the backdraft of combustion products from "naturally vented" appliances into the occupied space. This issue is of greater importance in new buildings with more air tight envelopes. To avoid the hazard, many modern combustion appliances utilize "direct venting" which draws combustion air directly from outdoors, instead of from the indoor environment.

Natural ventilation can also be achieved through the use of operable windows, this has largely been removed from most current architecture buildings due to the mechanical system continuously operating. The United States current strategy for ventilating buildings is to rely solely on mechanical ventilation. In Europe designers have experimented with design solutions that will allow for natural ventilation with minimal mechanical interference. These techniques include: building layout, facade construction, and materials used for inside finishes. European designers have also switched back to the use of operable windows to solve indoor air quality issues. "In the United States, the elimination of operable windows is one of the greatest losses in contemporary architecture."

Categories of Ventilation

- Mechanical ventilation refers to any system that uses mechanical means, such as a fan, to introduce subaerial air to a space. This includes positive pressure ventilation, exhaust ventilation, and balanced systems that use both supply and exhaust ventilation.
- Natural ventilation refers to intentionally designed passive methods of introducing subaerial to a space without the use of mechanical systems.
- Mixed mode ventilation (or hybrid ventilation) systems use both natural and mechanical processes.
- Infiltration is the uncontrolled flow of air from outdoors to indoors through leaks (unplanned openings) in a building envelope. When a building design relies on environmentally driven circumstantial infiltration to maintain indoor air quality, this flow has been referred to as adventitious ventilation.

Ventilation Rate Standards

The ventilation rate, for CII buildings, is normally expressed by the volumetric flowrate of subaerial air, introduced to the building. The typical units used are cubic feet per minute (CFM) or liters per second (L/s). The ventilation rate can also be expressed on a per person or per unit floor area basis, such as CFM/p or CFM/ft^2, or as *air changes per hour* (ACH).

Standards for Residential Buildings

For residential buildings, which mostly rely on infiltration for meeting their ventilation needs, a common ventilation rate measure is the air change rate (or air changes per hour): the hourly ventilation rate divided by the volume of the space (*I* or *ACH*; units of 1/h). During the winter, ACH may range from 0.50 to 0.41 in a tightly air-sealed house to 1.11 to 1.47 in a loosely air-sealed house.

ASHRAE now recommends ventilation rates dependent upon floor area, as a revision to the 62-2001 standard, in which the minimum ACH was 0.35, but no less than 15 CFM/person (7.1 L/s/person). As of 2003, the standard has been changed to 3 CFM/100 sq. ft. (15 l/s/100 sq. m.) plus 7.5 CFM/person (3.5 L/s/person).

In the UK, ventilation rate standards are specified in Part F of the Building Regulations.

Standards for Commercial Buildings

Ventilation Rate Procedure

Ventilation Rate Procedure is rate based on standard and prescribes the rate at which ventilation air must be delivered to a space and various means to condition that air. Air quality is assessed (through CO_2 measurement) and ventilation rates are mathematically derived using constants. Indoor Air Quality Procedure uses one or more guidelines for the specification of acceptable concentrations of certain contaminants in indoor air but does not prescribe ventilation rates or air treatment methods. This addresses both quantitative and subjective evaluations, and is based on the Ventilation Rate Procedure. It also accounts for potential contaminants that may have no measured limits, or for which no limits are not set (such as formaldehyde offgassing from carpet and furniture).

Table: Occupant Based Ventilation Rates, ANSI/ASHRAE Standard 62.1-2004.

IP Units	SI Units	Category	Examples
0 cfm/person	0 L/s/person	Spaces where ventilation requirements are primarily associated with building elements, not occupants.	Storage Rooms, Warehouses
5 cfm/person	2.5 L/s/person	Spaces occupied by adults, engaged in low levels of activity.	Office space
7.5 cfm/person	3.5 L/s/person	Spaces where occupants are engaged in higher levels of activity, but not strenuous, or activities generating more contaminants.	Retail spaces, lobbies
10 cfm/person	5 L/s/person	Spaces where occupants are engaged in more strenuous activity, but not exercise, or activities generating more contaminants.	Classrooms, school settings
20 cfm/person	10 L/s/person	Spaces where occupants are engaged in exercise, or activities generating many contaminants.	dance floors, exercise rooms

Table: Area-based ventilation rates, ANSI/ASHRAE Standard 62.1-2004.

IP Units	SI Units	Category	Examples
0.06 cfm/ft²	0.30 L/s/m²	Spaces where space contamination is normal, or similar to an office environment.	Conference rooms, lobbies
0.12 cfm/ft²	0.60 L/s/m²	Spaces where space contamination is significantly higher than an office environment.	Classrooms, museums
0.18 cfm/ft²	0.90 L/s/m²	Spaces where space contamination is even higher than the previous category.	Laboratories, art classrooms
0.30 cfm/ft²	1.5 L/s/m²	Specific spaces in sports or entertainment where contaminants are released.	Sports, entertainment
0.48 cfm/ft²	2.4 L/s/m²	Reserved for indoor swimming areas, where chemical concentrations are high.	Indoor swimming areas

The addition of occupant- and area-based ventilation rates found in the tables above often results in significantly reduced rates compared to the former standard. This is compensated in other sections of the standard which require that this minimum amount of air is actually delivered to the breathing zone of the individual occupant at all times. The total outdoor air intake of the ventilation system (in multiple-zone variable air volume

(VAV) systems) might therefore be similar to the airflow required by the 1989 standard. From 1999 to 2010, there was considerable development of the application protocol for ventilation rates. These advancements address occupant- and process-based ventilation rates, room ventilation effectiveness, and system ventilation effectiveness.

Natural Ventilation Strategies

The design of buildings that promote occupant health and well being requires clear understanding of the ways that ventilation airflow interacts with, dilutes, displaces or introduces pollutants within the occupied space. Although ventilation is an integral component to maintaining good indoor air quality, it may not be satisfactory alone. In scenarios where outdoor pollution would deteriorate indoor air quality, other treatment devices such as filtration may also be necessary. In kitchen ventilation systems, or for laboratory fume hoods, the design of effective effluent capture can be more important than the bulk amount of ventilation in a space. More generally, the way that an air distribution system causes ventilation to flow into and out of a space impacts the ability for a particular ventilation rate to remove internally generated pollutants. The ability for a system to remove pollution is described as its "ventilation effectiveness". However, the overall impacts of ventilation on indoor air quality can depend on more complex factors such as the sources of pollution, and the ways that activities and airflow interact to affect occupant exposure.

Techniques and architectural features used to ventilate buildings and structures naturally include, but are not limited to:

- Operable windows,
- Pressurised air pumps,
- Night purge ventilation,
- Clerestory windows and vented skylights,
- Building orientation,
- Wind capture façades.

Natural Ventilation

Natural ventilation harnesses naturally available forces to supply and remove air in an enclosed space. There are three types of natural ventilation occurring in buildings: wind driven ventilation, pressure-driven flows, and stack ventilation. The pressures generated by 'the stack effect' rely upon the buoyancy of heated or rising air. Wind driven ventilation relies upon the force of the prevailing wind to pull and push air through the enclosed space as well as through breaches in the building's envelope. Seoul University Professor Wonjun Kwon recently discovered a new way to ventilate large area of indoor space. The so-called "air pump" system uses pressure between inside and outside of rooms to push air out of a structure.

Almost all historic buildings were ventilated naturally. The technique was generally abandoned in larger US buildings during the late 20th century as the use of air conditioning became more widespread. However, with the advent of advanced Building Performance Simulation (BPS) software, improved Building Automation Systems (BAS), Leadership in Energy and Environmental Design (LEED) design requirements, and improved window manufacturing techniques; natural ventilation has made a resurgence in commercial buildings both globally and throughout the US.

The benefits of natural ventilation include:

- Improved Indoor air quality (IAQ),
- Energy savings,
- Reduction of greenhouse gas emissions,
- Occupant control,
- Reduction in occupant illness associated with Sick Building Syndrome,
- Increased worker productivity.

Mechanical Ventilation Strategies

Mechanical ventilation of buildings and structures can be achieved by use of the following techniques:

- Whole-house ventilation,
- Mixing ventilation,
- Displacement ventilation,
- Dedicated subaerial air supply.

Demand-controlled Ventilation (DCV)

Demand-controlled ventilation (DCV, also known as Demand Control Ventilation) makes it possible to maintain air quality while conserving energy. ASHRAE has determined that: "It is consistent with the ventilation rate procedure that demand control be permitted for use to reduce the total outdoor air supply during periods of less occupancy." In a DCV system, CO_2 sensors control the amount of ventilation. During peak occupancy, CO_2 levels rise, and the system adjusts to deliver the same amount of outdoor air as would be used by the ventilation-rate procedure. However, when spaces are less occupied, CO_2 levels reduce, and the system reduces ventilation to conserves energy. DCV is a well-established practice, and is required in high occupancy spaces by building energy standards such as ASHRAE 90.1.

Personalized Ventilation

Personalized ventilation is an air distribution strategy that allows individuals to control the amount of ventilation received. The approach deliver fresh air more directly to the breathing zone and aims to improve air quality of inhaled air. Personalized ventilation provides a much higher ventilation effectiveness than conventional mixing ventilation systems by displacing pollution from the breathing zone far less air volume. Beyond improved air quality benefits, the strategy can also improve occupant's thermal comfort, perceived air quality, and overall satisfaction with the indoor environment. Individual's preferences for temperature and air movement are not equal, and so traditional approaches to homogeneous environmental control have failed to achieve high occupant satisfaction. Techniques such as personalized ventilation facilitate control of a more diverse thermal environment that can improve thermal satisfaction for most occupants.

Local Exhaust Ventilation

Local exhaust ventilation addresses the issue of avoiding the contamination of indoor air by specific high-emission sources by capturing airborne contaminants before they are spread into the environment. This can include water vapor control, lavatory bioeffluent control, solvent vapors from industrial processes, and dust from wood- and metal-working machinery. Air can be exhausted through pressurized hoods or through the use of fans and pressurizing a specific area. A local exhaust system is composed of 5 basic parts:

- A hood that captures the contaminant at its source.
- Ducts for transporting the air.
- An air-cleaning device that removes/minimizes the contaminant.
- A fan that moves the air through the system.
- An exhaust stack through which the contaminated air is discharged.

In the UK, the use of LEV systems have regulations set out by the Health and Safety Executive (HSE) which are referred to as the Control of Substances Hazardous to Health (CoSHH). Under CoSHH, legislation is set out to protect users of LEV systems by ensuring that all equipment is tested at least every fourteen months to ensure the LEV systems are performing adequately. All parts of the system must be visually inspected and thoroughly tested and where any parts are found to be defective, the inspector must issue a red label to identify the defective part and the issue.

The owner of the LEV system must then have the defective parts repaired or replaced before the system can be used.

Smart Ventilation

Smart ventilation is a process to continually adjust the ventilation system in time, and optionally by location, to provide the desired IAQ benefits while minimizing energy consumption, utility bills and other non-IAQ costs (such as thermal discomfort or noise). A smart ventilation system adjusts ventilation rates in time or by location in a building to be responsive to one or more of the following: occupancy, outdoor thermal and air quality conditions, electricity grid needs, direct sensing of contaminants, operation of other air moving and air cleaning systems. In addition, smart ventilation systems can provide information to building owners, occupants, and managers on operational energy consumption and indoor air quality as well as signal when systems need maintenance or repair. Being responsive to occupancy means that a smart ventilation system can adjust ventilation depending on demand such as reducing ventilation if the building is unoccupied. Smart ventilation can time-shift ventilation to periods when a) indoor-outdoor temperature differences are smaller (and away from peak outdoor temperatures and humidity), b) when indoor-outdoor temperatures are appropriate for ventilative cooling, or c) when outdoor air quality is acceptable. Being responsive to electricity grid needs means providing flexibility to electricity demand (including direct signals from utilities) and integration with electric grid control strategies. Smart ventilation systems can have sensors to detect air flow, systems pressures or fan energy use in such a way that systems failures can be detected and repaired, as well as when system components need maintenance, such as filter replacement.

Ventilation and Combustion

Combustion (e.g., fireplace, gas heater, candle, oil lamp, etc.) consumes oxygen while producing carbon dioxide and other unhealthy gases and smoke, requiring ventilation air. An open chimney promotes infiltration (i.e. natural ventilation) because of the negative pressure change induced by the buoyant, warmer air leaving through the chimney. The warm air is typically replaced by heavier, cold air.

Ventilation in a structure is also needed for removing water vapor produced by respiration, burning, and cooking, and for removing odors. If water vapor is permitted to accumulate, it may damage the structure, insulation, or finishes. When operating, an air conditioner usually removes excess moisture from the air. A dehumidifier may also be appropriate for removing airborne moisture.

Smoking and Ventilation

ASHRAE standard 62 states that air removed from an area with environmental tobacco smoke shall not be recirculated into ETS-free air. A space with ETS requires more ventilation to achieve similar perceived air quality to that of a non-smoking environment.

The amount of ventilation in an ETS area is equal to the amount of ETS-free area plus the amount V, where:

$$V = DSD \times VA \times A/60E.$$

- V = Recommended extra flow rate in CFM (L/s).
- DSD = Design smoking density (estimated number of cigarettes smoked per hour per unit area).
- VA = Volume of ventilation air per cigarette for the room being designed (ft^3/cig).
- E = Contaminant removal effectiveness.

Air Quality Sampling

There are 5 main methods of sampling air quality:

Passive Monitoring

- Diffusion tubes absorb a specific pollutant from the ambient air - no power supply is needed.
- Diffusion tubes normally monitor for 2-4 weeks at a time.
- The tubes must be sent to a laboratory for analysis to see how much pollution they have detected.

Active (Semi Automatic) Sampling

- An analyser pulls the ambient air through a filter for a set period of time e.g one filter per day.
- The filters are then collected and sent to a laboratory for analysis to see how much pollution they have detected.

Automatic Point Monitoring

- Ambient air is pulled through an analyser which recognises the chosen gas and will calculate its concentration.
- Automatic sites monitor pollutants 24hours a day.

- Data is sent from the site directly to your computer which means it can be seen instantly.

Photochemical and Optical Sensor Systems

- These are portable monitoring tools that can continuously monitor a range of pollutants. The sensors are of low sensitivity and mostly suitable for identifying hotspots at roadsides and near point sources.
- Data can be downloaded to your computer and analysed.

Remote Optical/Long-path Monitoring

- This method of sampling detects pollution between a a light source and a detector which are placed separately at a site.
- Real time measurements can be taken with this type of sampling.
- Data can be sent from the analyser directly to your computer which means it can be seen instantly.

References

- May, jeffrey c. (2006). My office is killing me! : the sick building survival guide. Baltimore: the johns hopkins university press. Isbn 978-0-8018-8342-2
- What-is-air-quality: conserve-energy-future.com, Retrieved 14 July, 2020
- May, jeffrey c. (2001). My house is killing me! : the home guide for families with allergies and asthma. Baltimore: the johns hopkins university press. Isbn 978-0-8018-6730-9
- Air-quality-index: conserve-energy-future.com, Retrieved 17 May, 2020
- 7 million premature deaths annually linked to air pollution". Who. 25 march 2014. Retrieved 25 march 2014
- Brown, j. S.; gordon, t.; price, o.; asgharian, b. (2013). "thoracic and respirable particle definitions for human health risk assessment". Part fibre toxicol. 10: 12. Doi:10.1186/1743-8977-10-12. Pmc 3640939. Pmid 23575443
- All-about-effective-air-quality-monitoring, news: downtoearth.org.in, Retrieved 19 April, 2020
- Russell, colin a; hudson, john (2011-12-31). Early railway chemistry and its legacy. Royal society of chemistry. P. 67. Isbn 978-1-84973-326-7. Retrieved 2011-12-29

4

Soil Monitoring

Soil is a mixture of a number of organic matters, gases, minerals, liquids and organisms that support life. Soil monitoring includes the collection and analysis of soil, its quality and constituents. Soil erosion and soil pollution are a few of the focus areas of soil monitoring. This chapter discusses these aspects related to soil monitoring in detail.

SOIL

Loess field in Germany.

Soil is the mixture of minerals, organic matter, liquids, and gases covering most of the Earth's land surface and that serves, or has the ability to serve, as a medium for the growth of land plants. Although it may be covered by shallow water, if the water is too deep to support land plants (typically more than 2.5 meters), then the rock-covering mixture is not considered to be soil.

Soil is vitally important to all life on land. It supports rooted plants, provides a habitat and shelter to many animals, and it is the home to bacteria, fungi, and other microorganisms that recycle organic material for reuse by plants.

While the general concept of soil is well established, the definition of soil varies, according to the perspective of the discipline or occupation using soil as a resource.

Soil is among our most important natural resources because of its position in the landscape and its dynamic, physical, chemical, and biologic functions. It has been both used and misused. On the positive side, human creativity is expressed in using soil for agriculture, gardening and landscaping, utilizing peat as an energy source, producing fertilizers to replenish lost nutrients, employing soils as building materials (such as adobe and mudbrick), and transforming clay into eating and drinking vessels, storage containers, and works of art. On the other hand, anthropogenic activities have included fostering soil erosion and desertification through clear-cutting and overgrazing livestock, and contaminating soils by the dumping of industrial or household wastes.

The understanding of soil is incomplete. Despite the duration of humanity's dependence on and curiosity about soil, exploring the diversity and dynamic of this resource continues to yield fresh discoveries and insights. New avenues of soil research are compelled by our need to understand soil in the context of climate change and carbon sequestration. Our interest in maintaining the planet's biodiversity and in exploring past cultures has also stimulated renewed interest in achieving a more refined understanding of soil.

The earth's soil in general is sometimes referred to as comprising the pedosphere, which is positioned at the interface of the lithosphere with the biosphere, atmosphere, and hydrosphere. The scientific study of soil is called pedology or edaphology. Pedology is the study of soil in its natural setting, while edaphology is the study of soil in relation to soil-dependent uses.

Soil Components

Mineral Material

The majority of material in most soil is mineral. This consists of small grains broken off from the underlying rock or sometimes transported in from other areas by the action of water and wind. Larger mineral particles called sand and smaller particles called silt are the product of physical weathering, while even smaller particles called clay (a group of hydrous aluminium phyllosilicate minerals typically less than 2 micrometers in diameter) is generally the product of chemical weathering of silicate-bearing rocks. Clays are distinguished from other small particles present in soils such as silt by their small size, flake or layered shape, affinity for water and tendency toward high plasticity.

The mineral part of soil slowly releases nutrients that are needed by plants, such as potassium, calcium, and magnesium. Recently formed soil, for instance that formed from lava recently released from a volcano, is richer in nutrients and so is more fertile.

Organic Material

As plants and animals die and decay they return organic (carbon-bearing) material to the soil. Organic material tends to loosen up the soil and make it more productive for plant growth. Microorganisms, such as bacteria, fungi, and protists feed on the organic material and in the process release nutrients that can be reused by plants. The microorganisms themselves can form a significant part of the soil.

Water and Air

Soil almost always contains water and air in the spaces between the mineral and organic particles. Most soil organisms thrive best when the soil contains about equal volumes of water and air.

Soil Classification

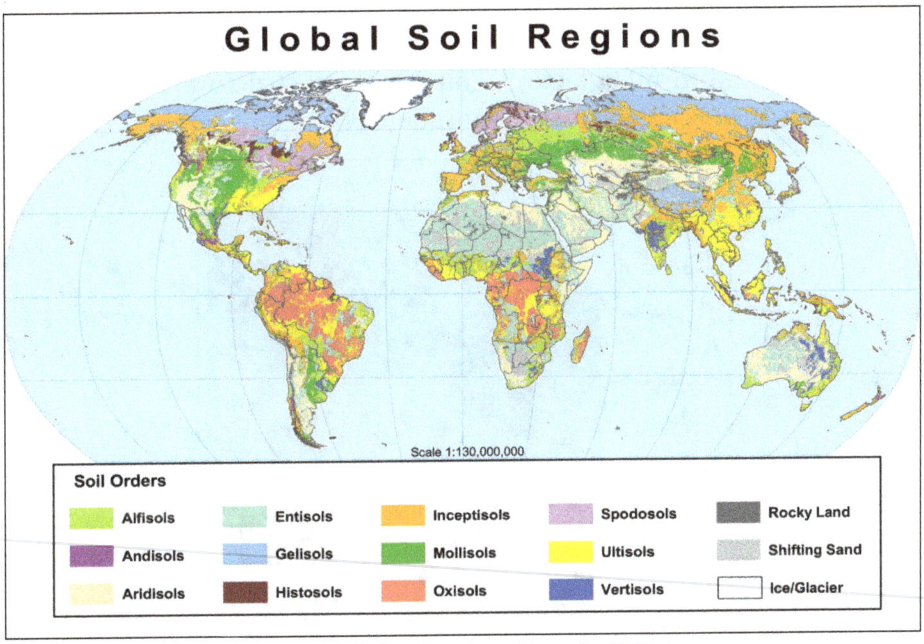

Map of global soil regions from the USDA.

The World Reference Base for Soil Resources (WRB) is the international standard soil classification system. Development of this system was coordinated by the International Soil Reference and Information Centre (ISRIC) and sponsored by the International Union of Soil Sciences (IUSS) and the Food and Agriculture Organization (FAO) via its Land and Water Development division. It replaces the previous FAO soil classification system.

The WRB borrows from modern soil classification concepts, including United States Department of Agriculture (USDA) soil taxonomy. The classification is based mainly on soil morphology as an expression of pedogenesis, the creation of soil. A major difference with USDA soil taxonomy is that soil climate is not part of the system, except in so far as climate influences soil profile characteristics.

The WRB structure is either nominal, giving unique names to soils or landscapes, or descriptive, naming soils by their characteristics such as red, hot, fat, or sandy. Soils are distinguished by obvious characteristics, such as physical appearance (e.g., color, texture, landscape position), performance (e.g., production capability, flooding), and accompanying vegetation. A vernacular distinction familiar to many is classifying texture as heavy or light. Light soils have lower clay content than heavy soils. They often drain better and dry out sooner, giving them a lighter color. Lighter soils, with their lower moisture content and better structure, take less effort to turn and cultivate. Contrary to popular belief light soils do not weigh less than heavy soils on an air dry basis nor do they have more porosity.

Soil Characteristics

Soils tend to develop an individualistic pattern of horizontal zonation under the influence of site specific soil-forming factors. Soil color, soil structure, and soil texture are especially important components of soil morphology.

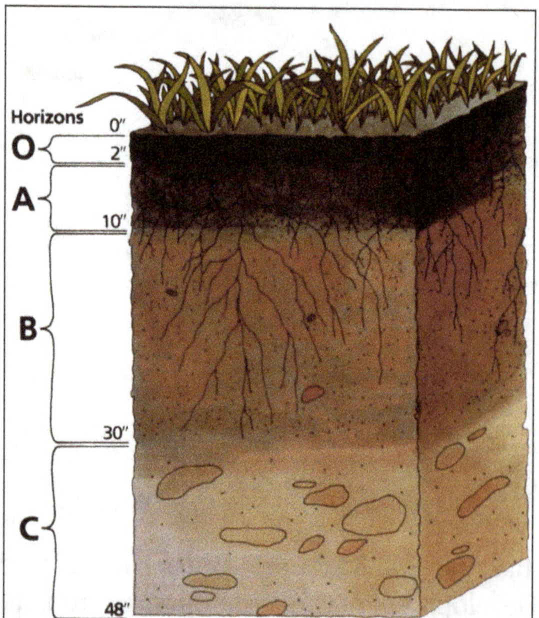

Soil horizons are formed by combined biological, chemical and physical alterations.

Soil color is the first impression one has when viewing soil. Striking colors and contrasting patterns are especially memorable. The Red River of the United States carries sediment eroded from extensive reddish soils like Port Silt Loam in Oklahoma. The Yellow River in China carries yellow sediment from eroding loessal soils. Mollisols in the Great Plains are darkened and enriched by organic matter. Podsols in boreal forests have highly contrasting layers due to acidity and leaching.

Soil color is primarily influenced by soil mineralogy. The extensive and various iron minerals in soil are responsible for an array of soil pigmentation. Color development and distribution of color within a soil profile result from chemical weathering, especially redox reactions. As the primary minerals in soil-parent material weather, the elements combine into new and colorful compounds. Iron forms secondary minerals with a yellow or red color; organic matter decomposes into black and brown compounds; and manganese forms black mineral deposits. These pigments give soil its various colors and patterns and are further affected by environmental factors. Aerobic conditions produce uniform or gradual color changes, while reducing environments result in disrupted color flow with complex, mottled patterns and points of color concentration.

Soil structure is the arrangement of soil particles into aggregates. These may have various shapes, sizes and degrees of development or expression. Soil structure influences aeration, water movement, erosion resistance, and root penetration. Observing structure gives clues to texture, chemical and mineralogical conditions, organic content, biological activity, and past use, or abuse.

Surface soil structure is the primary component of tilth. Where soil mineral particles are both separated and bridged by organic-matter-breakdown products and soil-biota exudates, it makes the soil easy to work. Cultivation, earthworms, frost action, and rodents mix the soil. This activity decreases the size of the peds to form a granular (or crumb) structure. This structure allows for good porosity and easy movement of air and water. The combination of ease in tillage, good moisture and air-handling capabilities, good structure for planting and germination are definitive of good tilth.

Soil texture refers to sand, silt and clay composition in combination with gravel and larger-material content. Clay content is particularly influential on soil behavior due to a high retention capacity for nutrients and water. Due to superior aggregation, clay soils resist wind and water erosion better than silty and sandy soils. In medium-textured soils, clay can tend to move downward through the soil profile to accumulate as illuvium in the subsoil. The lighter-textured, surface soils are more responsive to management inputs, but also more vulnerable to erosion and contamination.

Texture influences many physical aspects of soil behavior. Available water capacity increases with silt and, more importantly, clay content. Nutrient-retention capacity tends to follow the same relationship. Plant growth, and many uses which rely on soil, tends to favor medium-textured soils, such as loam and sandy loam. A balance in air and water-handling characteristics within medium-textured soils are largely responsible for this.

Soil and its Environment

Soil and Plants

Soil is necessary for almost all land plants to survive and grow. A sprouting seed sends into the soil roots, which absorb water and dissolved minerals that the new plant needs for its growth. As the plant grows its root system expands though the soil and serves to support it. The soil stores water from rain and snow, allowing plants to have a continuous supply and helping to prevent destructive flooding. As plants lose their leaves, and when the plants themselves die, organic material is returned to the soil, which builds up and enriches the soil. In many environments, plants also protect the soil from erosion by holding it in place with their roots and sheltering it from the effects of wind and rain.

Soil and Animals

Since all animals depend on plants for their food, directly or indirectly, all land animals depend on the soil that makes plant life on land possible. Some animals, such as earthworms and moles, live all their lives in the soil itself, while others, such as ground squirrels and most ants, live in homes dug into the soil, protecting them from predators and from heat and cold and bad weather. Other animals, such as pigs, dig into the soil to find food, and others, such as beavers and some birds, use soil as a building material. Bison, elephants, and some other large animals cover themselves with soil for protection against sun and insects.

The wastes of animals enrich the soil and their digging mixes and loosens it; both of these activities benefit the further growth of plants. In New York State (United States), it is estimated the woodchucks turn over 1.6 million tons (1.63 million metric tons) of soil each year.

Soil in Nature

Soil formation processes never stop and soil is always changing. The long periods over which change occurs and the multiple influences of change mean that simple soils are rare. While soil can achieve relative stability in properties for extended periods of time, the soil life cycle ultimately ends in soil conditions that leave it vulnerable to erosion. Little of the soil continuum of the earth is older than Tertiary and most are no older than Pleistocene. Despite the inevitability of soil retrogression and degradation, most soil cycles are long and productive. How the soil "life" cycle proceeds is influenced by at least five classic soil forming factors: regional climate, biotic potential, topography, parent material, and the passage of time.

An example of soil development from bare rock occurs on recent lava flows in warm regions under heavy and very frequent rainfall. In such climates, plants become established very quickly on basaltic lava, even though there is very little organic material. The plants are supported by the porous rock becoming filled with nutrient-bearing water, for example, carrying dissolved bird droppings or guano. The developing plant roots themselves gradually breaks up the porous lava and organic matter soon accumulates but, even before it does, the predominantly porous broken lava in which the plant roots grow can be considered a soil.

Sample of an aerial photo from a published soil survey.

Most of our knowledge of soil in nature comes from soil survey efforts. Soil survey, or soil mapping, is the process of determining the soil types or other properties of the soil cover over a landscape, and mapping them for others to understand and use. It relies heavily on distinguishing the individual influences of the five classic soil forming factors. This effort draws upon geomorphology, physical geography, and analysis of vegetation and land-use patterns. Primary data for the soil survey are acquired by field sampling and supported by remote sensing.

Geologists have a particular interest in the patterns of soil on the surface of the earth. Soil texture, color and chemistry often reflect the underlying geologic parent material and soil types often change at geologic unit boundaries. Geologists classify surface soils using the 1938 USDA soil taxonomy, but use the current version of USDA soil taxonomy to classify the buried soils that make up the paleopedological record. Buried paleosols mark previous land surfaces and record climatic conditions from previous eras. Geologists use this paleopedological record to understand the ecological relationships in past ecosystems. According to the theory of biorhexistasy, prolonged conditions conducive to forming deep, weathered soils result in increasing ocean salinity and the formation of limestone.

Geologists and pedologists use soil profile features to establish the duration of surface stability in the context of geologic faults or slope stability. An offset subsoil horizon indicates rupture during soil formation and the degree of subsequent subsoil formation is relied upon to establish time since rupture.

Soil examined in shovel test pits is used by archaeologists for relative dating based on stratigraphy (as opposed to absolute dating). What is considered most typical is to use soil profile features to determine the maximum reasonable pit depth than needs to be examined for archaeological evidence in the interest of cultural resources management.

Soils altered or formed by man (anthropic and anthropogenic soils) are also of interest to archaeologists. An example is Terra preta do Indio, found in the Amazon river basin.

Soil Quality

Testing soil fertility.

Soil quality is a measure of the condition of soil relative to the requirements of one or more biotic species and or to any human need or purpose. According to the United States Department of Agriculture Natural Resources Conservation Service, "Soil quality is the capacity of a specific kind of soil to function, within natural or managed ecosystem boundaries, to sustain plant and animal productivity, maintain or enhance water and air quality, and support human health and habitation. The European Commission's Joint Research Centre proposed a definition, stating that "Soil quality is an account of the soil's ability to provide ecosystem and social services through its capacities to perform its functions under changing conditions."

Soil quality reflects how well a soil performs the functions of maintaining biodiversity and productivity, partitioning water and solute flow, filtering and buffering, nutrient cycling, and providing support for plants and other structures. Soil management has a major impact on soil quality.

Soil quality in agricultural terms is measured on a scale of soil value (*Bodenwertzahl*) in Germany. Soil quality relates to soil functions. Unlike water or air, for which established standards have been set, soil quality is difficult to define or quantify. Soil quality can be evaluated using the Soil Management Assessment Framework.

SOIL MONITORING

Collecting a soil sample in Mexico for pathogen testing.

Soil monitoring involves the collection and/or analysis of soil and its associated quality, constituents, and physical status to determine or guarantee its fitness for use. Soil faces many threats, including compaction, contamination, organic material loss, biodiversity loss, slope stability issues, erosion, salinization, and acidification. Soil monitoring helps characterize these and other potential risks to the soil, surrounding environments, animal health, and human health.

Assessing these and other risks to soil can be challenging due to a variety of factors, including soil's heterogeneity and complexity, scarcity of toxicity data, lack of understanding of a contaminant's fate, and variability in levels of soil screening. This requires a risk assessment approach and analysis techniques that prioritize environmental protection, risk reduction, and, if necessary, remediation methods. Soil monitoring plays a significant role in that risk assessment, not only aiding in the identification of at-risk and affected areas but also in the establishment of base background values of soil.

Soil monitoring has historically focused on more classical conditions and contaminants, including toxic elements (e.g., mercury, lead, and arsenic) and persistent organic pollutants (POPs). Historically, testing for these and other aspects of soil, however, has had its own set of challenges, as sampling in most cases is of a destructive in nature, requiring multiple samples over time. Additionally, procedural and analytical errors may be introduced due to variability among references and methods, particularly over time. However, as analytical techniques evolve and new knowledge about ecological processes and contaminant effects disseminate, the focus of monitoring will likely broaden over time and the quality of monitoring will continue to improve.

Soil Sampling

The two primary types of soil sampling are grab sampling and composite sampling. Grab sampling involves the collection of an individual sample at a specific time and place, while composite sampling involves the collection of a homogenized mixture of multiple individual samples at either a specific place over different times or multiple locations at a specific time. Soil sampling may occur both at shallow ground levels or deep in the ground, with collection methods varying by level collected from. Scoops, augers, core barrel and solid-tube samplers, and other tools are used shallow, whereas split-tube, solid-tube, or hydraulic methods may be used in deep ground.

Monitoring Programs

Soil Contamination Monitoring

A portable X-ray fluorescence (XRF) analyzer can be used in the field for testing soils for metal contamination.

Soil contamination monitoring helps researchers identify patterns and trends in contaminant deposition, movement, and effect. Human-based pressures such as tourism, industrial activity, urban sprawl, construction work, and inadequate agriculture/forestry practices can contribute to and make worse soil contamination and lead to the soil becoming unfit for its intended use. Both inorganic and organic pollutants may make their way to the soil, having a wide variety of detrimental effects. Soil contamination monitoring is therefor important to identify risk areas, set baselines, and identify contaminated zones for remediation. Monitoring efforts may range from local farms to nationwide efforts, such as those made by China in the late 2000s, providing details such as the nature of contaminants, their quantity, effects, concentration patterns, and remediation feasibility. Monitoring and analytical equipment will ideally will have high response times, high levels of resolution and automation, and a certain degree of self-sufficiency. Chemical techniques may be used to measure toxic elements and POPs using chromatography and spectrometry, geophysical techniques may assess physical properties of large terrains, and biological techniques may use specific organisms to gauge not only contaminant level but also byproducts of contaminant biodegradation.

These techniques and others are increasingly becoming more efficient, and laboratory instrumentation is becoming more precise, resulting in more meaningful monitoring outcomes.

Soil Erosion Monitoring

Soil erosion monitoring helps researchers identify patterns and trends in soil and sediment movement. Monitoring programs have varied over the years, from long-term academic research on university plots to reconnaissance-based surveys of biogeoclimatic areas. In most methods, however, the general focus is on identifying and measuring all the dominant erosion processes in a given area. Additionally, soil erosion monitoring may attempt to quantify the effects of erosion on crop productivity, though challenging "because of the many complexities in the relationship between soils and plants and their management under a variable climate."

Soil Salinity Monitoring

Soil salinity monitoring helps researchers identify patterns and trends in soil salt content. Both the natural process of seawater intrusion and the human-induced processes of inappropriate soil and water management can lead to salinity problems in soil, with up to one billion hectares of land affected globally. Salinity monitoring at the local level may look closely at the root zone to gauge salinity impact and develop management options, where at the regional and national level salinity monitoring may help with identifying areas at-risk and aiding policymakers in tackling the issue before it spreads. The monitoring process itself may be performed using technologies such as remote sensing and geographic information systems (GIS) to identify salinity via greenness, brightness, and whiteness at the surface level. Direct analysis of soil up close, including the use of electromagnetic induction techniques, may also be used to monitor soil salinity.

SOIL POLLUTION

Soil contamination or soil pollution as part of land degradation is caused by the presence of xenobiotics (human-made) chemicals or other alteration in the natural soil environment. It is typically caused by industrial activity, agricultural chemicals or improper disposal of waste. The most common chemicals involved are petroleum hydrocarbons, polynuclear aromatic hydrocarbons (such as naphthalene and benzo(a) pyrene), solvents, pesticides, lead, and other heavy metals. Contamination is correlated with the degree of industrialization and intensity of chemical substance. The concern over soil contamination stems primarily from health risks, from direct contact with the contaminated soil, vapours from the contaminants, and from secondary contamination of water supplies within and underlying the soil. Mapping of contaminated soil sites and the resulting cleanups are time consuming and expensive tasks, requiring extensive amounts of geology, hydrology, chemistry, computer modeling skills, and GIS in Environmental Contamination, as well as an appreciation of the history of industrial chemistry.

Excavation showing soil contamination at a disused gasworks in England.

In North America and Western Europe the extent of contaminated land is best known, with many of countries in these areas having a legal framework to identify and deal with this environmental problem. Developing countries tend to be less tightly regulated despite some of them having undergone significant industrialization.

Uses

Soil pollution can be caused by the following:

- Microplastics.
- Oil spills.
- Mining and activities by other heavy industries.
- Accidental spills may happen during activities, etc.
- Corrosion of underground storage tanks (including piping used to transmit the contents).
- Acid rain.
- Intensive farming.
- Agrochemicals, such as pesticides, herbicides and fertilizers.
- Petrochemicals.

- Industrial accidents.
- Road debris.
- Drainage of contaminated surface water into the soil.
- Ammunitions, chemical agents, and other agents of war.
- Waste disposal.
    - Oil and fuel dumping.
    - Nuclear wastes.
    - Direct discharge of industrial wastes to the soil.
    - Discharge of sewage.
    - Landfill and illegal dumping.
    - Coal ash.
    - Electronic waste.

The most common chemicals involved are petroleum hydrocarbons, solvents, pesticides, lead, and other heavy metals.

Any activity that leads to other forms of soil degradation (erosion, compaction, etc.) may indirectly worsen the contamination effects in that soil remediation becomes more tedious.

Historical deposition of coal ash used for residential, commercial, and industrial heating, as well as for industrial processes such as ore smelting, were a common source of contamination in areas that were industrialized before about 1960. Coal naturally concentrates lead and zinc during its formation, as well as other heavy metals to a lesser degree. When the coal is burned, most of these metals become concentrated in the ash (the principal exception being mercury). Coal ash and slag may contain sufficient lead to qualify as a "characteristic hazardous waste", defined in the USA as containing more than 5 mg/l of extractable lead using the TCLP procedure. In addition to lead, coal ash typically contains variable but significant concentrations of polynuclear aromatic hydrocarbons (PAHs; e.g., benzo(a)anthracene, benzo(b)fluoranthene, benzo(k)fluoranthene, benzo(a)pyrene, indeno(cd)pyrene, phenanthrene, anthracene, and others). These PAHs are known human carcinogens and the acceptable concentrations of them in soil are typically around 1 mg/kg. Coal ash and slag can be recognised by the presence of off-white grains in soil, gray heterogeneous soil, or (coal slag) bubbly, vesicular pebble-sized grains.

Treated sewage sludge, known in the industry as biosolids, has become controversial as a "fertilizer". As it is the byproduct of sewage treatment, it generally contains more contaminants such as organisms, pesticides, and heavy metals than other soil.

In the European Union, the Urban Waste Water Treatment Directive allows sewage sludge to be sprayed onto land. The volume is expected to double to 185,000 tons of dry solids in 2005. This has good agricultural properties due to the high nitrogen and phosphate content. In 1990/1991, 13% wet weight was sprayed onto 0.13% of the land; however, this is expected to rise 15 fold by 2005. Advocates say there is a need to control this so that pathogenic microorganisms do not get into water courses and to ensure that there is no accumulation of heavy metals in the top soil.

Pesticides and Herbicides

A pesticide is a substance used to kill a pest. A pesticide may be a chemical substance, biological agent (such as a virus or bacteria), antimicrobial, disinfectant or device used against any pest. Pests include insects, plant pathogens, weeds, mollusks, birds, mammals, fish, nematodes (roundworms) and microbes that compete with humans for food, destroy property, spread or are a vector for disease or cause a nuisance. Although there are benefits to the use of pesticides, there are also drawbacks, such as potential toxicity to humans and other organisms.

Herbicides are used to kill weeds, especially on pavements and railways. They are similar to auxins and most are biodegradable by soil bacteria. However, one group derived from trinitrotoluene (2:4 D and 2:4:5 T) have the impurity dioxin, which is very toxic and causes fatality even in low concentrations. Another herbicide is Paraquat. It is highly toxic but it rapidly degrades in soil due to the action of bacteria and does not kill soil fauna.

Insecticides are used to rid farms of pests which damage crops. The insects damage not only standing crops but also stored ones and in the tropics it is reckoned that one third of the total production is lost during food storage. As with fungicides, the first insecticides used in the nineteenth century were inorganic e.g. Paris Green and other compounds of arsenic. Nicotine has also been used since the late eighteenth century.

There are now two main groups of synthetic insecticides:

1. Organochlorines include DDT, Aldrin, Dieldrin and BHC. They are cheap to produce, potent and persistent. DDT was used on a massive scale from the 1930s, with a peak of 72,000 tonnes used 1970. Then usage fell as the harmful environmental effects were realized. It was found worldwide in fish and birds and was even discovered in the snow in the Antarctic. It is only slightly soluble in water but is very soluble in the bloodstream. It affects the nervous and endocrine systems and causes the eggshells of birds to lack calcium causing them to be easily breakable. It is thought to be responsible for the decline of the numbers of birds of prey like ospreys and peregrine falcons in the 1950s – they are now recovering. As well as increased concentration via the food chain,

it is known to enter via permeable membranes, so fish get it through their gills. As it has low water solubility, it tends to stay at the water surface, so organisms that live there are most affected. DDT found in fish that formed part of the human food chain caused concern, but the levels found in the liver, kidney and brain tissues was less than 1 ppm and in fat was 10 ppm, which was below the level likely to cause harm. However, DDT was banned in the UK and the United States to stop the further buildup of it in the food chain. U.S. manufacturers continued to sell DDT to developing countries, who could not afford the expensive replacement chemicals and who did not have such stringent regulations governing the use of pesticides.

2. Organophosphates, e.g. parathion, methyl parathion and about 40 other insecticides are available nationally. Parathion is highly toxic, methyl-parathion is less so and Malathion is generally considered safe as it has low toxicity and is rapidly broken down in the mammalian liver. This group works by preventing normal nerve transmission as cholinesterase is prevented from breaking down the transmitter substance acetylcholine, resulting in uncontrolled muscle movements.

Agents of War

The disposal of munitions, and a lack of care in manufacture of munitions caused by the urgency of production, can contaminate soil for extended periods. There is little published evidence on this type of contamination largely because of restrictions placed by governments of many countries on the publication of material related to war effort. However, mustard gas stored during World War II has contaminated some sites for up to 50 years and the testing of Anthrax as a potential biological weapon contaminated the whole island of Gruinard.

Health Effects

Contaminated or polluted soil directly affects human health through direct contact with soil or via inhalation of soil contaminants which have vaporized; potentially greater threats are posed by the infiltration of soil contamination into groundwater aquifers used for human consumption, sometimes in areas apparently far removed from any apparent source of above ground contamination. This tends to result in the development of pollution-related diseases.

Health consequences from exposure to soil contamination vary greatly depending on pollutant type, pathway of attack and vulnerability of the exposed population. Chronic exposure to chromium, lead and other metals, petroleum, solvents, and many pesticide and herbicide formulations can be carcinogenic, can cause congenital disorders, or can cause other chronic health conditions. Industrial or man-made concentrations of naturally occurring substances, such as nitrate and ammonia associated with livestock manure from agricultural operations, have also been identified as health hazards in soil and groundwater.

Chronic exposure to benzene at sufficient concentrations is known to be associated with higher incidence of leukemia. Mercury and cyclodienes are known to induce higher incidences of kidney damage and some irreversible diseases. PCBs and cyclodienes are linked to liver toxicity. Organophosphates and carbonates can induce a chain of responses leading to neuromuscular blockage. Many chlorinated solvents induce liver changes, kidney changes and depression of the central nervous system. There is an entire spectrum of further health effects such as headache, nausea, fatigue, eye irritation and skin rash for the above cited and other chemicals. At sufficient dosages a large number of soil contaminants can cause death by exposure via direct contact, inhalation or ingestion of contaminants in groundwater contaminated through soil.

The Scottish Government has commissioned the Institute of Occupational Medicine to undertake a review of methods to assess risk to human health from contaminated land. The overall aim of the project is to work up guidance that should be useful to Scottish Local Authorities in assessing whether sites represent a significant possibility of significant harm (SPOSH) to human health. It is envisaged that the output of the project will be a short document providing high level guidance on health risk assessment with reference to existing published guidance and methodologies that have been identified as being particularly relevant and helpful. The project will examine how policy guidelines have been developed for determining the acceptability of risks to human health and propose an approach for assessing what constitutes unacceptable risk in line with the criteria for SPOSH as defined in the legislation and the Scottish Statutory Guidance.

Ecosystem Effects

Not unexpectedly, soil contaminants can have significant deleterious consequences for ecosystems. There are radical soil chemistry changes which can arise from the presence of many hazardous chemicals even at low concentration of the contaminant species. These changes can manifest in the alteration of metabolism of endemic microorganisms and arthropods resident in a given soil environment. The result can be virtual eradication of some of the primary food chain, which in turn could have major consequences for predator or consumer species. Even if the chemical effect on lower life forms is small, the lower pyramid levels of the food chain may ingest alien chemicals, which normally become more concentrated for each consuming rung of the food chain. Many of these effects are now well known, such as the concentration of persistent DDT materials for avian consumers, leading to weakening of egg shells, increased chick mortality and potential extinction of species.

Effects occur to agricultural lands which have certain types of soil contamination. Contaminants typically alter plant metabolism, often causing a reduction in crop yields. This has a secondary effect upon soil conservation, since the languishing crops cannot shield the Earth's soil from erosion. Some of these chemical contaminants have long half-lives and in other cases derivative chemicals are formed from decay of primary soil contaminants.

Cleanup Options

Cleanup or environmental remediation is analyzed by environmental scientists who utilize field measurement of soil chemicals and also apply computer models (GIS in Environmental Contamination) for analyzing transport and fate of soil chemicals. Various technologies have been developed for remediation of oil-contaminated soil and sediments There are several principal strategies for remediation:

- Excavate soil and take it to a disposal site away from ready pathways for human or sensitive ecosystem contact. This technique also applies to dredging of bay muds containing toxins.
- Aeration of soils at the contaminated site (with attendant risk of creating air pollution).
- Thermal remediation by introduction of heat to raise subsurface temperatures sufficiently high to volatize chemical contaminants out of the soil for vapor extraction. Technologies include ISTD, electrical resistance heating (ERH), and ET-DSP.
- Bioremediation, involving microbial digestion of certain organic chemicals. Techniques used in bioremediation include landfarming, biostimulation and bioaugmentating soil biota with commercially available microflora.
- Extraction of groundwater or soil vapor with an active electromechanical system, with subsequent stripping of the contaminants from the extract.
- Containment of the soil contaminants (such as by capping or paving over in place).
- Phytoremediation, or using plants (such as willow) to extract heavy metals.
- Mycoremediation, or using fungus to metabolize contaminants and accumulate heavy metals.
- Remediation of oil contaminated sediments with self-collapsing air microbubbles.
- Surfactant leaching.

Soil Pollutant

Soil pollutants can be included into two main groups, the organic pollutants (OPs) and the inorganic pollutants (IPs). One of the best examples of the latter are the potentially toxic elements (PTEs). Both PTEs and OPs can have natural or anthropogenic origins and the majority of polluted soils in the world contain complex mixtures of each or from both. Although these two types of pollutants highly differ in their behavior in soils. This are mainly related to the nondegradability of PTEs in relation to OPs, which can be decomposed by living organisms. Thus, PTEs may experience increased or decreased availability over time, depending on their form when deposited in soil, and changes in physicochemical conditions during their accumulation in soil. On the other hand, the availability of OPs tends to decrease over time and is closely related to their degradation susceptibility, usually resulting in decomposition into simple units or functional groups of lowest toxicity, but not necessarily.

Potentially Toxic Elements (PTEs)

PTEs is a general collective term applied to the group of metals and metalloids with an atomic density greater than 4 g cm^{-3}, which toxicity for living organisms are well reported. This group of elements is originally derived from natural sources through the weathering of rocks and minerals, or other geological events (e.g., volcanic emissions). However, anthropogenic activities are gaining special relevance as the main sources of PTEs pollution in the soil, since the rates of PTEs generation via manmade cycles are more rapid relative to natural ones. Major anthropogenic sources of PTEs for the different environmental compartments are presented in table.

Table: Anthropogenic sources of several PTEs, established by the US Environmental Protection Agency based on potential hazard to human health, and introduced into the environment via effluent discharges, emissions, runoff or waste disposal.

Industrial process	As	Be	Cd	Cr	Cu	Hg	Ni	Pb	Sb	Se	Tl	Zn
Alloys	■		■	■	■		■	■	■	■		■
Batteries and electro/chemical cells		■				■	■	■		■		■
Biocides (agriculture, antifouling)	■					■						
Ceramics and glass	■						■	■		■	■	
Chemicals, pharmaceuticals, dental	■		■	■	■	■		■	■	■		■
Coatings (anticorrosives)			■	■				■				■
Electrical equipment and apparatus		■			■			■				
Fertilizers	■		■	■	■	■	■	■				■
Fossil fuel combustion (electricity)	■		■			■		■	■	■		
Mining, smelting, metallurgy	■	■	■	■	■	■	■	■	■	■	■	■
Nuclear reactor (moderator, absorber)		■	■									
Paints and pigments	■		■	■	■		■	■	■	■		■
Petroleum refining	■			■	■		■	■				■
Pipes, sheets, machinery					■			■				
Plastics			■					■				■
Pulp and paper				■	■	■	■	■				
Rubber										■		■
Semiconductors, superconductors	■							■			■	
Tanning and textiles	■			■	■							
Wood preservative treatment	■			■								

PTEs are considered priority pollutants, since their nondegradability and persistence in environment for long periods allows their transfer from the contamination sources (e.g., mining or industrial areas) to other locations where direct exposure of living organisms may be more favored. The new conditions of the receiving system may induce changes in chemical form (species) and render it more (bio)available.

In soil system, PTEs can occur in the soil solution under ionic, molecular, chelated, and colloidal forms, conferring them high mobility, or associated with the solid fractions. Thus, from the ready to less available PTEs forms are (1) exchangeable ions in mineral or organic particles; (2) complexed or chelated to organic colloids; (3) sorbed to inorganic constituents; (4) incorporated in supergenic phases as (oxy)hydroxides, clay minerals, or insoluble salts; and (5) fixed in crystal lattice of the minerals. Even when PTEs are not dissolved, their mobility in soil is always dependent of an aqueous phase, as well as the intrinsic characteristics of each element. Both will determine the ease of release (dissolution), or sorption onto surfaces of soil components, according to the physicochemical conditions of the medium, being the pH and ion interactions determining factors on this. In addition, the mobility of PTEs are highly dependent on the source, and usually PTEs from anthropogenic inputs are more mobile than from pedogenic or geogenic ones.

Therefore, the knowledge of PTEs support phases are a crucial step to assess their migration and availability to other environmental compartments, highlighting the bioavailability for plants, animals, and humans, since concentrations in these bioavailable fractions are not necessarily proportional to the total concentration of PTEs. In general, the free ionic species are the most bioavailable forms of PTEs and good indicators of toxicity. However, this is not a rule since there are cases, such as the organic forms of Hg (methylmercury), that are more toxic than the free ionic species. In addition, the toxicity of some PTEs change according to their oxidation states. A good example is the Cr(VI), which is toxic and carcinogen for living organisms and also very mobile in soils, whereas Cr(III) is not toxic to plants and is necessary in animal nutrition. Thus, reductants materials in soils, as the case of organic matter, are very important in the reduction of Cr(VI) to more stable Cr(III) complexes.

Organic Pollutants (OPs)

Although OPs can occur naturally, associated to some volcanic emission, forest fires, or related to fossil fuels, in fact, the greatest harmful impacts are associated with the manmade production of a huge number of chemical compounds. For many of these chemicals, the environmental and health impacts are not even known, while for others the risks are well known and result in prohibition of their use as the typical example of DDT.

The most important factors regarding OPs toxicity are the persistence, the solubility (in water or organic solvents), the volatilization and the by-products resulting from their (bio)degradation. table summarizes the main properties that control the behavior of OPs in various soil processes.

Table: Main OPs properties, soil parameters, and environmental conditions with reference to the most important processes that influence their behavior in soil systems.

Parameters	Main processes that determine the fate of OPs		
	Nonionic sorption	Ionic sorption	Degradation
Soil			
Clay contents		■	
Oxides contents (DCB soluble Fe+Al)		■	
Content of organic matter (OM)	■		■
OM acid dissociation constant (pKa)		■	
OM composition	■		■
Carbonate content		■	■
Microbial mass			■
Soil structure	■	■	■
Environmental conditions			
pH		■	■
Eh/redox			■
Temperature			■
EC		■	
Composition of soluble anions and cations		■	
Organic pollutant			
Electrical charge		■	
Hydrophobicity (KOC/W)	■	■	
Lipophilicity			■
Molecular weight	■		■
Functional groups	■	■	■

The OPs that are being deserve more attention in last decades are the persistent organic pollutants (POPs), which, as the name indicates, have long half-lives in environment due to their resistance to biological, chemical, and photolytic degradation. The hydrophobic that they present causing their retention in the particle soil fraction, particularly in organic matter due to being lipophilic compounds. In addition, the great affinity for lipids allows the accumulation in food chains by storage in fatty tissues of organisms. The semivolatile character of these pollutants also enable them to move long distances in the atmosphere, which cause a high spatial distribution and away from the source.

Considering the high diversity of POPs, they can be grouped into two main categories considering the production purpose or the result (by-product) of industrial activity. As main chemical synthesis products are that for industrial uses (e.g., PCBs) and that ones for agrochemicals (pesticides) uses (e.g., DDT and its metabolites, toxaphene, chlordane). As examples of combustion or industrial by-products of chemical synthesis are dioxins/furans (PCDDs/Fs) and polycyclic aromatic hydrocarbons (PAHs), which can be direct contaminants as they do not have a specific application purpose. PAHs are considered priority pollutants by US EPA and EU and their distribution in the environment, as well as the potential risks to human health (some of them are probable human carcinogens) have been the focus of much attention. These compounds, which inputs can be natural or anthropogenic, have soil as the most important environmental compartment sink. This may result from their hydrophobicity, which make them readily adsorbable by soil particles, particularly SOM, and difficult of being degraded. On the other hand, more polar POPs are phenols compounds, making easier their transport in soils, since they can exist as dissolved in soil solution, sorbed to soil particles, or polymerized in humic compounds. For this reason, they are easily degraded under aerobic conditions.

SOIL EROSION

An actively eroding rill on an intensively-farmed field in eastern Germany.

Soil erosion is the displacement of the upper layer of soil, it is one form of soil degradation. This natural process is caused by the dynamic activity of erosive agents, that is, water, ice (glaciers), snow, air (wind), plants, animals, and humans. In accordance with these agents, erosion is sometimes divided into water erosion, glacial erosion, snow erosion, wind (aeolean) erosion, zoogenic erosion and anthropogenic erosion. Soil erosion may be a slow process that continues relatively unnoticed, or it may occur at an alarming rate causing a serious loss of topsoil. The loss of soil from farmland may be reflected in reduced crop production potential, lower surface water quality and damaged drainage networks.

Human activities have increased by 10–50 times the rate at which erosion is occurring globally. Excessive (or accelerated) erosion causes both "on-site" and "off-site" problems. On-site impacts include decreases in agricultural productivity and (on natural landscapes) ecological collapse, both because of loss of the nutrient-rich upper soil layers. In some cases, the eventual end result is desertification. Off-site effects include sedimentation of waterways and eutrophication of water bodies, as well as sediment-related damage to roads and houses. Water and wind erosion are the two primary causes of land degradation; combined, they are responsible for about 84% of the global extent of degraded land, making excessive erosion one of the most significant environmental problems worldwide.

Intensive agriculture, deforestation, roads, anthropogenic climate change and urban sprawl are amongst the most significant human activities in regard to their effect on stimulating erosion. However, there are many prevention and remediation practices that can curtail or limit erosion of vulnerable soils.

Physical Processes

Rainfall and Surface Runoff

Rainfall, and the surface runoff which may result from rainfall, produces four main types of soil erosion: splash erosion, sheet erosion, rill erosion, and gully erosion. Splash erosion is generally seen as the first and least severe stage in the soil erosion process, which is followed by sheet erosion, then rill erosion and finally gully erosion (the most severe of the four).

Soil and water being splashed by the impact of a single raindrop.

In splash erosion, the impact of a falling raindrop creates a small crater in the soil, ejecting soil particles. The distance these soil particles travel can be as much as 0.6 m (two feet) vertically and 1.5 m (five feet) horizontally on level ground.

If the soil is saturated, or if the rainfall rate is greater than the rate at which water can infiltrate into the soil, surface runoff occurs. If the runoff has sufficient flow energy, it will transport loosened soil particles (sediment) down the slope. *Sheet erosion* is the transport of loosened soil particles by overland flow.

A spoil tip covered in rills and gullies due to erosion processes caused by rainfall.

Rill erosion refers to the development of small, ephemeral concentrated flow paths which function as both sediment source and sediment delivery systems for erosion on hillslopes. Generally, where water erosion rates on disturbed upland areas are greatest, rills are active. Flow depths in rills are typically of the order of a few centimeters (about an inch) or less and along-channel slopes may be quite steep. This means that rills exhibit hydraulic physics very different from water flowing through the deeper, wider channels of streams and rivers.

Gully erosion occurs when runoff water accumulates and rapidly flows in narrow channels during or immediately after heavy rains or melting snow, removing soil to a considerable depth.

Rivers and Streams

Valley or *stream erosion* occurs with continued water flow along a linear feature. The erosion is both downward, deepening the valley, and headward, extending the valley into the hillside, creating head cuts and steep banks. In the earliest stage of stream erosion, the erosive activity is dominantly vertical, the valleys have a typical V cross-section and the stream gradient is relatively steep. When some base level is reached, the erosive activity switches to lateral erosion, which widens the valley floor and creates a narrow floodplain. The stream gradient becomes nearly flat, and lateral deposition of sediments becomes important as the stream meanders across the valley floor. In all stages of stream erosion, by far the most erosion occurs during times of flood, when more and faster-moving water is available to carry a larger sediment load. In such processes, it is not the water alone that erodes: suspended abrasive particles, pebbles and boulders can also act erosively as they traverse a surface, in a process known as *traction*.

Dobbingstone Burn, Scotland—This photo illustrates two different types of erosion affecting the same place. Valley erosion is occurring due to the flow of the stream, and the boulders and stones (and much of the soil) that are lying on the edges are glacial till that was left behind as ice age glaciers flowed over the terrain.

Bank erosion is the wearing away of the banks of a stream or river. This is distinguished from changes on the bed of the watercourse, which is referred to as *scour*. Erosion and changes in the form of river banks may be measured by inserting metal rods into the bank and marking the position of the bank surface along the rods at different times.

Thermal erosion is the result of melting and weakening permafrost due to moving water. It can occur both along rivers and at the coast. Rapid river channel migration observed in the Lena River of Siberia is due to thermal erosion, as these portions of the banks are composed of permafrost-cemented non-cohesive materials. Much of this erosion occurs as the weakened banks fail in large slumps. Thermal erosion also affects the Arctic coast, where wave action and near-shore temperatures combine to undercut permafrost bluffs along the shoreline and cause them to fail. Annual erosion rates along a 100-kilometre (62-mile) segment of the Beaufort Sea shoreline averaged 5.6 metres (18 feet) per year from 1955 to 2002.

Floods

At extremely high flows, kolks, or vortices are formed by large volumes of rapidly rushing water. Kolks cause extreme local erosion, plucking bedrock and creating pothole-type geographical features called Rock-cut basins. Examples can be seen in the flood regions result from glacial Lake Missoula, which created the channeled scablands in the Columbia Basin region of eastern Washington.

Wind Erosion

Wind erosion is a major geomorphological force, especially in arid and semi-arid regions. It is also a major source of land degradation, evaporation, desertification, harmful airborne dust, and crop damage—especially after being increased far above natural rates by human activities such as deforestation, urbanization, and agriculture.

Árbol de Piedra, a rock formation in the Altiplano, Bolivia sculpted by wind erosion.

Wind erosion is of two primary varieties: *deflation*, where the wind picks up and carries away loose particles; and *abrasion*, where surfaces are worn down as they are struck by airborne particles carried by wind. Deflation is divided into three categories: (1) *surface creep*, where larger, heavier particles slide or roll along the ground; (2) *saltation*, where particles are lifted a short height into the air, and bounce and saltate across the surface of the soil; and (3) *suspension*, where very small and light particles are lifted into the air by the wind, and are often carried for long distances. Saltation is responsible for the majority (50–70%) of wind erosion, followed by suspension (30–40%), and then surface creep (5–25%). Silty soils tend to be the most affected by wind erosion; silt particles are relatively easily detached and carried away.

Wind erosion is much more severe in arid areas and during times of drought. For example, in the Great Plains, it is estimated that soil loss due to wind erosion can be as much as 6100 times greater in drought years than in wet years.

Mass Movement

Mass movement is the downward and outward movement of rock and sediments on a sloped surface, mainly due to the force of gravity.

Mass movement is an important part of the erosional process, and is often the first stage in the breakdown and transport of weathered materials in mountainous areas. It moves material from higher elevations to lower elevations where other eroding agents such as streams and glaciers can then pick up the material and move it to even lower elevations. Mass-movement processes are always occurring continuously on all slopes; some mass-movement processes act very slowly; others occur very suddenly, often with disastrous results. Any perceptible down-slope movement of rock or sediment is often referred to in general terms as a landslide. However, landslides can be classified in a much more detailed way that reflects the mechanisms responsible for the movement and the velocity at which the movement occurs. One of the visible topographical manifestations of a very slow form of such activity is a scree slope.

Wadi in Makhtesh Ramon, Israel, showing gravity collapse erosion on its banks.

Slumping happens on steep hillsides, occurring along distinct fracture zones, often within materials like clay that, once released, may move quite rapidly downhill. They will often show a spoon-shaped isostatic depression, in which the material has begun to slide downhill. In some cases, the slump is caused by water beneath the slope weakening it. In many cases it is simply the result of poor engineering along highways where it is a regular occurrence.

Surface creep is the slow movement of soil and rock debris by gravity which is usually not perceptible except through extended observation. However, the term can also describe the rolling of dislodged soil particles 0.5 to 1.0 mm (0.02 to 0.04 in) in diameter by wind along the soil surface.

Factors affecting Soil Erosion

Climate

The amount and intensity of precipitation is the main climatic factor governing soil erosion by water. The relationship is particularly strong if heavy rainfall occurs at times when, or in locations where, the soil's surface is not well protected by vegetation. This might be during periods when agricultural activities leave the soil bare, or in semi-arid regions where vegetation is naturally sparse. Wind erosion requires strong winds, particularly during times of drought when vegetation is sparse and soil is dry (and so is more erodible). Other climatic factors such as average temperature and temperature range may also affect erosion, via their effects on vegetation and soil properties. In general, given similar vegetation and ecosystems, areas with more precipitation (especially high-intensity rainfall), more wind, or more storms are expected to have more erosion.

In some areas of the world (e.g. the mid-western USA), rainfall intensity is the primary determinant of erosivity, with higher intensity rainfall generally resulting in more soil erosion by water. The size and velocity of rain drops is also an important factor. Larger and higher-velocity rain drops have greater kinetic energy, and thus their impact will displace soil particles by larger distances than smaller, slower-moving rain drops.

In other regions of the world (e.g. western Europe), runoff and erosion result from relatively low intensities of stratiform rainfall falling onto previously saturated soil. In such situations, rainfall amount rather than intensity is the main factor determining the severity of soil erosion by water.

Soil Structure and Composition

Erosional gully in unconsolidated Dead Sea (Israel) sediments along the southwestern shore. This gully was excavated by floods from the Judean Mountains in less than a year.

The composition, moisture, and compaction of soil are all major factors in determining the erosivity of rainfall. Sediments containing more clay tend to be more resistant to erosion than those with sand or silt, because the clay helps bind soil particles together. Soil containing high levels of organic materials are often more resistant to erosion, because the organic materials coagulate soil colloids and create a stronger, more stable soil structure. The amount of water present in the soil before the precipitation also plays an important role, because it sets limits on the amount of water that can be absorbed by the soil (and hence prevented from flowing on the surface as erosive runoff). Wet, saturated soils will not be able to absorb as much rain water, leading to higher levels of surface runoff and thus higher erosivity for a given volume of rainfall. Soil compaction also affects the permeability of the soil to water, and hence the amount of water that flows away as runoff. More compacted soils will have a larger amount of surface runoff than less compacted soils.

Vegetative Cover

Vegetation acts as an interface between the atmosphere and the soil. It increases the permeability of the soil to rainwater, thus decreasing runoff. It shelters the soil from winds, which results in decreased wind erosion, as well as advantageous changes in microclimate. The roots of the plants bind the soil together, and interweave with other roots, forming a more solid mass that is less susceptible to both water and wind erosion. The removal of vegetation increases the rate of surface erosion.

Topography

The topography of the land determines the velocity at which surface runoff will flow, which in turn determines the erosivity of the runoff. Longer, steeper slopes (especially those without adequate vegetative cover) are more susceptible to very high rates of erosion during heavy rains than shorter, less steep slopes. Steeper terrain is also more prone to mudslides, landslides, and other forms of gravitational erosion processes.

Human Activities that Increase Soil Erosion

Agricultural Practices

Tilled farmland such as this is very susceptible to erosion from rainfall, due to the destruction of vegetative cover and the loosening of the soil during plowing.

Unsustainable agricultural practices increase rates of erosion by one to two orders of magnitude over the natural rate and far exceed replacement by soil production. The tillage of agricultural lands, which breaks up soil into finer particles, is one of the primary factors. The problem has been exacerbated in modern times, due to mechanized agricultural equipment that allows for deep plowing, which severely increases the amount of soil that is available for transport by water erosion. Others include mono-cropping, farming on steep slopes, pesticide and chemical fertilizer usage (which kill organisms that bind soil together), row-cropping, and the use of surface irrigation. A complex overall situation with respect to defining nutrient losses from soils, could arise as a result of the size selective nature of soil erosion events. Loss of total phosphorus, for instance, in the finer eroded fraction is greater relative to the whole soil. Extrapolating this evidence to predict subsequent behaviour within receiving aquatic systems, the reason is that this more easily transported material may support a lower solution P concentration compared to coarser sized fractions. Tillage also increases wind erosion rates, by dehydrating the soil and breaking it up into smaller particles that can be picked up by the wind. Exacerbating this is the fact that most of the trees are generally removed from agricultural fields, allowing winds to have long, open runs to travel over at higher speeds. Heavy grazing reduces vegetative cover and causes severe soil compaction, both of which increase erosion rates.

Deforestation

In an undisturbed forest, the mineral soil is protected by a layer of *leaf litter* and an *humus* that cover the forest floor. These two layers form a protective mat over the soil that absorbs the impact of rain drops. They are porous and highly permeable to rainfall, and allow rainwater to slow percolate into the soil below, instead of flowing over the surface as runoff. The roots of the trees and plants hold together soil particles, preventing them from being washed away. The vegetative cover acts to reduce the velocity of the raindrops that strike the foliage and stems before hitting the ground, reducing their kinetic energy. However it is the forest floor, more than the canopy, that prevents surface erosion. The terminal velocity of rain drops is reached in about 8 metres (26 feet). Because forest canopies are usually higher than this, rain drops can often regain terminal velocity even after striking the canopy. However, the intact forest floor, with its layers of leaf litter and organic matter, is still able to absorb the impact of the rainfall.

In this clearcut, almost all of the vegetation has been stripped from surface of steep slopes, in an area with very heavy rains. Severe erosion occurs in cases such as this, causing stream sedimentation and the loss of nutrient rich topsoil.

Deforestation causes increased erosion rates due to exposure of mineral soil by removing the humus and litter layers from the soil surface, removing the vegetative cover that binds soil together, and causing heavy soil compaction from logging equipment. Once trees have been removed by fire or logging, infiltration rates become high and erosion low to the degree the forest floor remains intact. Severe fires can lead to significant further erosion if followed by heavy rainfall.

Globally one of the largest contributors to erosive soil loss in the year 2006 is the slash and burn treatment of tropical forests. In a number of regions of the earth, entire sectors of a country have been rendered unproductive. For example, on the Madagascar high central plateau, comprising approximately ten percent of that country's land area, virtually the entire landscape is sterile of vegetation, with gully erosive furrows typically in excess of 50 metres (160 ft) deep and 1 kilometre (0.6 miles) wide. Shifting cultivation is a farming system which sometimes incorporates the slash and burn method in some regions of the world. This degrades the soil and causes the soil to become less and less fertile.

Roads and Urbanization

Urbanization has major effects on erosion processes—first by denuding the land of vegetative cover, altering drainage patterns, and compacting the soil during construction; and next by covering the land in an impermeable layer of asphalt or concrete that increases the amount of surface runoff and increases surface wind speeds. Much of the sediment carried in runoff from urban areas (especially roads) is highly contaminated with fuel, oil, and other chemicals. This increased runoff, in addition to eroding and degrading the land that it flows over, also causes major disruption to surrounding watersheds by altering the volume and rate of water that flows through them, and filling them with chemically polluted sedimentation. The increased flow of water through local waterways also causes a large increase in the rate of bank erosion.

Climate Change

The warmer atmospheric temperatures observed over the past decades are expected to lead to a more vigorous hydrological cycle, including more extreme rainfall events. The rise in sea levels that has occurred as a result of climate change has also greatly increased coastal erosion rates.

Studies on soil erosion suggest that increased rainfall amounts and intensities will lead to greater rates of soil erosion. Thus, if rainfall amounts and intensities increase in many parts of the world as expected, erosion will also increase, unless amelioration measures are taken. Soil erosion rates are expected to change in response to changes in climate for a variety of reasons. The most direct is the change in the erosive power of rainfall. Other reasons include: a) changes in plant canopy caused by shifts in plant biomass production associated with moisture regime; b) changes in litter cover on the ground caused by changes in both plant residue decomposition rates driven by temperature and moisture dependent soil microbial activity as well as plant biomass production rates; c) changes in soil moisture due to shifting precipitation regimes and evapo-transpiration rates, which changes infiltration and runoff ratios; d) soil erodibility changes due to decrease in soil organic matter concentrations in soils that lead to a soil structure that is more susceptible to erosion and increased runoff due to increased soil surface sealing and crusting; e) a shift of winter precipitation from non-erosive snow to erosive rainfall due to increasing winter temperatures; f) melting of permafrost, which induces an erodible soil state from a previously non-erodible one; and g) shifts in land use made necessary to accommodate new climatic regimes.

Studies by Pruski and Nearing indicated that, other factors such as land use unconsidered, it is reasonable to expect approximately a 1.7% change in soil erosion for each 1% change in total precipitation under climate change. In recent studies, there are predicted increases of rainfall erosivity by 17% in the United States and by 18% in Europe.

Global Environmental Effects

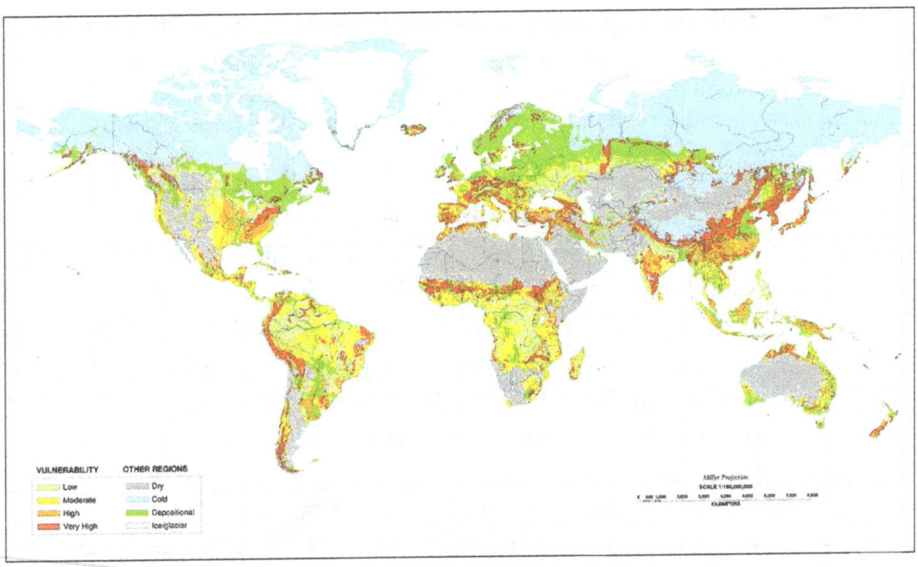

World map indicating areas that are vulnerable to high rates of water erosion.

Due to the severity of its ecological effects, and the scale on which it is occurring, erosion constitutes one of the most significant global environmental problems we face today.

Land Degradation

Water and wind erosion are now the two primary causes of land degradation; combined, they are responsible for 84% of degraded acreage.

Each year, about 75 billion tons of soil is eroded from the land—a rate that is about 13–40 times as fast as the natural rate of erosion. Approximately 40% of the world's agricultural land is seriously degraded. According to the United Nations, an area of fertile soil the size of Ukraine is lost every year because of drought, deforestation and climate change. In Africa, if current trends of soil degradation continue, the continent might be able to feed just 25% of its population by 2025, according to UNU's Ghana-based Institute for Natural Resources in Africa.

During the 17th and 18th centuries, Easter Island experienced severe erosion due to deforestation and unsustainable agricultural practices. The resulting loss of topsoil ultimately led to ecological collapse, causing mass starvation and the complete disintegration of the Easter Island civilization.

Recent modeling developments have quantified rainfall erosivity at global scale using high temporal resolution(<30 min) and high fidelity rainfall recordings. The results is an extensive global data collection effort produced the Global Rainfall Erosivity Database (GloREDa) which includes rainfall erosivity for 3,625 stations and covers 63 countries. This first ever Global Rainfall Erosivity Database was used to develop a global erosivity map at 30 arc-seconds(~1 km) based on sophisticated geostatistical process. According to a new study published in Nature Communications, almost 36 billion tons of soil is lost every year due to water, and deforestation and other changes in land use make the problem worse. The study investigates global soil erosion dynamics by means of high-resolution spatially distributed modelling (ca. 250 × 250 m cell size). The geo-statistical approach allows, for the first time, the thorough incorporation into a global soil erosion model of land use and changes in land use, the extent, types, spatial distribution of global croplands and the effects of different regional cropping systems.

The loss of soil fertility due to erosion is further problematic because the response is often to apply chemical fertilizers, which leads to further water and soil pollution, rather than to allow the land to regenerate.

Sedimentation of Aquatic Ecosystems

Soil erosion (especially from agricultural activity) is considered to be the leading global cause of diffuse water pollution, due to the effects of the excess sediments flowing into the world's waterways. The sediments themselves act as pollutants, as well as being carriers for other pollutants, such as attached pesticide molecules or heavy metals.

The effect of increased sediments loads on aquatic ecosystems can be catastrophic. Silt can smother the spawning beds of fish, by filling in the space between gravel on the stream bed. It also reduces their food supply, and causes major respiratory issues for them as sediment enters their gills. The biodiversity of aquatic plant and algal life is reduced, and invertebrates are also unable to survive and reproduce. While the sedimentation event itself might be relatively short-lived, the ecological disruption caused by the mass die off often persists long into the future.

One of the most serious and long-running water erosion problems worldwide is in the People's Republic of China, on the middle reaches of the Yellow River and the upper reaches of the Yangtze River. From the Yellow River, over 1.6 billion tons of sediment flows into the ocean each year. The sediment originates primarily from water erosion in the Loess Plateau region of the northwest.

Airborne Dust Pollution

Soil particles picked up during wind erosion of soil are a major source of air pollution, in the form of airborne particulates—"dust". These airborne soil particles are often contaminated with toxic chemicals such as pesticides or petroleum fuels, posing ecological and public health hazards when they later land, or are inhaled/ingested.

Dust from erosion acts to suppress rainfall and changes the sky color from blue to white, which leads to an increase in red sunsets. Dust events have been linked to a decline in the health of coral reefs across the Caribbean and Florida, primarily since the 1970s. Similar dust plumes originate in the Gobi desert, which combined with pollutants, spread large distances downwind, or eastward, into North America.

Monitoring, Measuring and Modeling Soil Erosion

Terracing is an ancient technique that can significantly slow the rate of water erosion on cultivated slopes.

Monitoring and modeling of erosion processes can help people better understand the causes of soil erosion, make predictions of erosion under a range of possible conditions, and plan the implementation of preventative and restorative strategies for erosion. However, the complexity of erosion processes and the number of scientific disciplines that must be considered to understand and model them (e.g. climatology, hydrology, geology, soil science, agriculture, chemistry, physics, etc.) makes accurate modelling challenging. Erosion models are also non-linear, which makes them difficult to work with numerically, and makes it difficult or impossible to scale up to making predictions about large areas from data collected by sampling smaller plots.

The most commonly used model for predicting soil loss from water erosion is the *Universal Soil Loss Equation (USLE)*. This was developed in the 1960s and 1970s. It estimates the average annual soil loss *A* on a plot-sized area as:

$$A = RKLSCP$$

where *R* is the *rainfall erosivity factor*, *K* is the *soil erodibility factor*, *L* and *S* are topographic factors representing length and slope, *C* is the cover and management factor and *P* is the support practices factor.

Despite the USLE's plot-scale spatial focus, the model has often been used to estimate soil erosion on much larger areas, such as watersheds or even whole continents. For example, RUSLE has recently been used to quantify soil erosion across the whole of Europe. One major problem is that the USLE cannot simulate gully erosion, and so erosion from gullies is ignored in any USLE-based assessment of erosion. Yet erosion from gullies can be a substantial proportion (10–80%) of total erosion on cultivated and grazed land.

During the 50 years since the introduction of the USLE, many other soil erosion models have been developed. But because of the complexity of soil erosion and its constituent processes, all erosion models can give unsatisfactory results when validated i.e. when model predictions are compared with real-world measurements of erosion. Thus new soil erosion models continue to be developed. Some of these remain USLE-based, e.g. the G2 model. Other soil erosion models have largely (e.g. the Water Erosion Prediction Project model) or wholly (e.g. the Rangeland Hydrology and Erosion Model) abandoned usage of USLE elements.

Prevention and Remediation

A windbreak (the row of trees) planted next to an agricultural field, acting as a shield against strong winds. This reduces the effects of wind erosion, and provides many other benefits.

The most effective known method for erosion prevention is to increase vegetative cover on the land, which helps prevent both wind and water erosion. Terracing is an extremely effective means of erosion control, which has been practiced for thousands of years by people all over the world. Windbreaks (also called shelterbelts) are rows of trees and shrubs that are planted along the edges of agricultural fields, to shield the fields against winds. In addition to significantly reducing wind erosion, windbreaks provide many other benefits such as improved microclimates for crops (which are sheltered from the dehydrating and otherwise damaging effects of wind), habitat for beneficial bird species, carbon sequestration, and aesthetic improvements to the agricultural landscape. Traditional planting methods, such as mixed-cropping (instead of monocropping) and crop rotation have also been shown to significantly reduce erosion rates. Crop residues play a role in the mitigation of erosion, because they reduce the impact of raindrops breaking up the soil particles. There is a higher potential for erosion when producing potatoes than when growing cereals, or oilseed crops. Forages have a fibrous root system, which helps combat erosion by anchoring the plants to the top layer of the soil, and covering the entirety of the field, as it is a non-row crop. In tropical coastal systems, properties of mangroves have been examined as a potential means to reduce soil erosion. Their complex root structures are known to help reduce wave damage from storms and flood impacts while binding and building soils. These roots can slow down water flow, leading to the deposition of sediments and reduced erosion rates. However, in order to maintain sediment balance, adequate mangrove forest width needs to be present. Strip cropping consists of growing erosion-permitting and erosion-resisting crops alternately in strips across the slope and on the approximate contour. In this system, the crops such as groundnut and soya bean are included as erosion-resisting crops and jowar and bajra as erosion-permitting crops.

Erosion Prediction

There are dozens of erosion prediction models. Some models focus on long-term (natural or geological) erosion, as a component of landscape evolution. However, many erosion models were developed to quantify the effects of accelerated soil erosion i.e. soil erosion as influenced by human activity.

Most soil erosion models consider only soil erosion by water, however a few aim to predict wind erosion. Models which consider tillage erosion are rare. Also soil erosion models have been more commonly developed for use on agricultural landscapes, rather than on naturally vegetated areas (such as rangeland or forests). A few erosion models focus on erosion on mined areas.

The aim of the majority of soil erosion models is to predict average rates (often an annual average rate) of soil loss from an area such as a plot, a field or a catchment/watershed under various land management techniques. Some erosion models are purely statistical, others more mechanistic (or physically based). Two of the more widely used soil erosion models in North America are the Revised Universal Soil Loss Equation (RUSLE) and the Water Erosion Prediction Project erosion model (WEPP). Much of the mineland erosion literature is solely focused on fitting or improving RUSLE parameters. Few soil erosion models consider gully erosion, mostly due to difficulties in modelling these large erosional features.

Several studies have evaluated the ability of soil erosion models to realistically predict measured rates of erosion, mainly on agricultural landscapes. There is often a wide discrepancy between predicted and observed erosion rates. Thus soil erosion models are still better as research tools than as public policy and regulatory instruments or for prescriptive design measures for constructed landforms. However soil erosion models may provide useful guidance for the design engineer if adequately calibrated and verified for local conditions and if the design accounts for the uncertainty.

Most erosion modelling is applied to existing sites of known topography and material properties to guide land management activities. Designers of constructed landforms, however, have considerable control over the topography, cover soil placement, initial revegetation, and to a lesser extent the substrate properties – flexibility that is generally uneconomical for farmers and ranchers and most users of erosion models. On the other hand, miners have little input into post-closure land use practices and management.

Methods to estimate erosion rates include:

- Purely statistical models.
- Subjectively determined erosion rates using expert judgement combined with a database of erosion rates of natural and reclaimed sites (natural and industrial analogs).
- Surveying of existing erosional or depositional features of known age (or as determined by dating of deposits) to determine average erosion rates. Analysis of historical aerial photographs is often employed.
- Site-specific empirical models that relate slope, watershed size, and rainfall.
- Empirical and semi-empirical or deterministic models based on laboratory and field flume measurements of erosion under simulated rainfall or flow conditions.
- Physically based gully erosion models.
- Landform and landscape scale models, often GIS-based, that apply erosion mechanics or statistical relationships to predict changes in topography and erosion rates.

- Sediment-budget models based on watershed monitoring.

Erosion Control

Hydroseeding in the United Kingdom.

Erosion control is the practice of preventing or controlling wind or water erosion in agriculture, land development, coastal areas, river banks and construction. Effective erosion controls handle surface runoff and are important techniques in preventing water pollution, soil loss, wildlife habitat loss and human property loss.

Terraces, conservation tillage, and conservation buffers save soil and improve water quality on this Iowa farm.

Usage

Erosion controls are used in natural areas, agricultural settings or urban environments. In urban areas erosion controls are often part of stormwater runoff management programs required by local governments. The controls often involve the creation of a physical barrier, such as vegetation or rock, to absorb some of the energy of the wind or water that is causing the erosion. They also involve building and maintaining storm drains. On construction sites they are often implemented in conjunction with sediment controls such as sediment basins and silt fences.

Bank erosion is a natural process: without it, rivers would not meander and change course. However, land management patterns that change the hydrograph and/or vegetation cover can act to increase or decrease channel migration rates. In many places, whether or not the banks are unstable due to human activities, people try to keep a river in a single place. This can be done for environmental reclamation or to prevent a river from changing course into land that is being used by people. One way that this is done is by placing riprap or gabions along the bank.

Examples of erosion control methods include:

- Buffer strip.
- Cellular confinement systems.
- Crop rotation.
- Conservation tillage.
- Contour bunding.
- Contour plowing.
- Cover crops.
- Fiber rolls.
- Gabions.
- Hydroseeding.
- Level spreaders.
- Mulching.
- Perennial crops.
- Plasticulture.
- Polyacrylamide (as a coagulant).
- Reforestation.
- Riparian buffer.
- Riprap.

- Strip farming.
- Sand fence.
- Vegetated waterway (bioswale).
- Terracing.
- Wattle (construction).
- Windbreaks.

Mathematical Modeling

Since the 1920s and 1930s scientists have been creating mathematical models for understanding the mechanisms of soil erosion and resulting sediment surface runoff, including an early paper by Albert Einstein applying Baer's law. These models have addressed both gully and sheet erosion. Earliest models were a simple set of linked equations which could be employed by manual calculation. By the 1970s the models had expanded to complex computer models addressing nonpoint source pollution with thousands of lines of computer code. The more complex models were able to address nuances in micrometeorology, soil particle size distributions and micro-terrain variation.

SOIL SALINITY

Soil salinity is the salt content in the soil; the process of increasing the salt content is known as salinizations. Salts occur naturally within soils and water. Salination can be caused by natural processes such as mineral weathering or by the gradual withdrawal of an ocean. It can also come about through artificial processes such as irrigation and road salt.

Visibly salt-affected soils on rangeland in Colorado. Salts dissolved from the soil accumulate at the soil surface and are deposited on the ground and at the base of the fence post.

Saline incrustation in a PVC irrigation pipe from Brazil.

Natural Occurrence

Salts are a natural component in soils and water. The ions responsible for salination are: Na^+, K^+, Ca^{2+}, Mg^{2+} and Cl^-. As the Na^+ (sodium) predominates, soils can become sodic. Sodic soils present particular challenges because they tend to have very poor structure which limits or prevents water infiltration and drainage.

Over long periods of time, as soil minerals weather and release salts, these salts are flushed or leached out of the soil by drainage water in areas with sufficient precipitation. In addition to mineral weathering, salts are also deposited via dust and precipitation. In dry regions salts may accumulate, leading to naturally saline soils. This is the case, for example, in large parts of Australia. Human practices can increase the salinity of soils by the addition of salts in irrigation water. Proper irrigation management can prevent salt accumulation by providing adequate drainage water to leach added salts from the soil. Disrupting drainage patterns that provide leaching can also result in salt accumulations. An example of this occurred in Egypt in 1970 when the Aswan High Dam was built. The change in the level of ground water before the construction had enabled soil erosion, which led to high concentration of salts in the water table. After the construction, the continuous high level of the water table led to the salination of the arable land.

Dry Land Salinity

Salinity in drylands can occur when the water table is between two and three metres from the surface of the soil. The salts from the groundwater are raised by capillary action to the surface of the soil. This occurs when groundwater is saline (which is true in many areas), and is favored by land use practices allowing more rainwater to enter the aquifer than it could accommodate. For example, the clearing of trees for agriculture is a major reason for dryland salinity in some areas, since deep rooting of trees has been replaced by shallow rooting of annual crops.

Salinity due to Irrigation

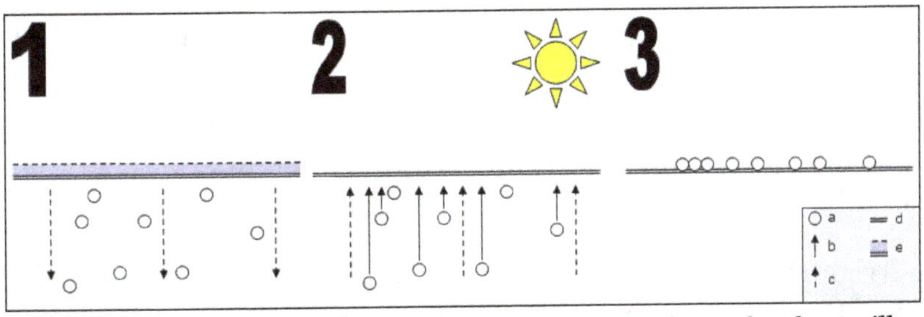

Rain or irrigation, in the absence of leaching, can bring salts to the surface by capillary action.

Salinity from irrigation can occur over time wherever irrigation occurs, since almost all water (even natural rainfall) contains some dissolved salts. When the plants use the water, the salts are left behind in the soil and eventually begin to accumulate. Since soil salinity makes it more difficult for plants to absorb soil moisture, these salts must be leached out of the plant root zone by applying additional water. This water in excess of plant needs is called the leaching fraction. Salination from irrigation water is also greatly increased by poor drainage and use of saline water for irrigating agricultural crops.

Salinity in urban areas often results from the combination of irrigation and groundwater processes. Irrigation is also now common in cities (gardens and recreation areas).

Consequences of Salinity

The consequences of salinity are:

- Detrimental effects on plant growth and yield.
- Damage to infrastructure (roads, bricks, corrosion of pipes and cables).
- Reduction of water quality for users, sedimentation problems, increased leaching of metals, especially copper, cadmium, manganese and zinc.
- Soil erosion ultimately, when crops are too strongly affected by the amounts of salts.
- More energy required to desalinate.

Salinity is an important land degradation problem. Soil salinity can be reduced by leaching soluble salts out of soil with excess irrigation water. Soil salinity control involves watertable control and flushing in combination with tile drainage or another form of subsurface drainage. A comprehensive treatment of soil salinity is available from the United Nations Food and Agriculture Organization.

Salt Tolerance of Crops

High levels of soil salinity can be tolerated if salt-tolerant plants are grown. Sensitive crops lose their vigor already in slightly saline soils, most crops are negatively affected by (moderately) saline soils, and only salinity-resistant crops thrive in severely saline soils. The University of Wyoming and the Government of Alberta report data on the salt tolerance of plants.

Field data in irrigated lands, under farmers' conditions, are scarce, especially in developing countries. However, some on-farm surveys have been made in Egypt, India, and Pakistan. Some examples are shown in the following gallery, with crops arranged from sensitive to very tolerant.

Graphs of crop yield and soil salinity in farmers' fields ordered by increasing salt tolerance.

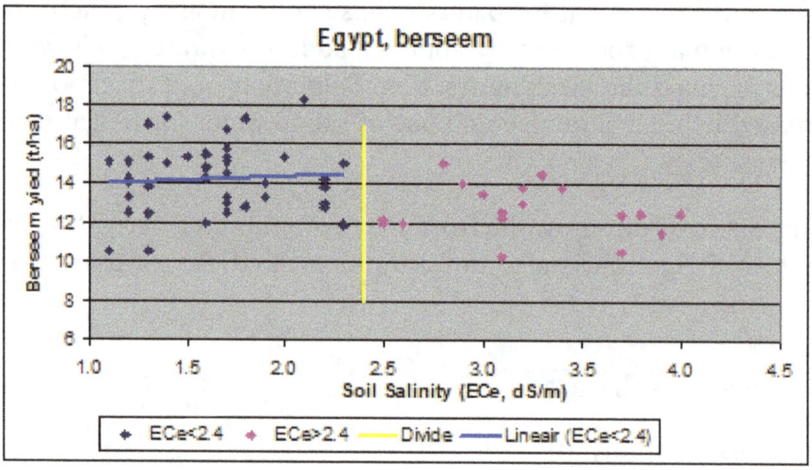

Berseem (clover), cultivated in Egypt's Nile Delta, is a salt-sensitive crop and tolerates an ECe value up to 2.4 dS/m, whereafter yields start to decline.

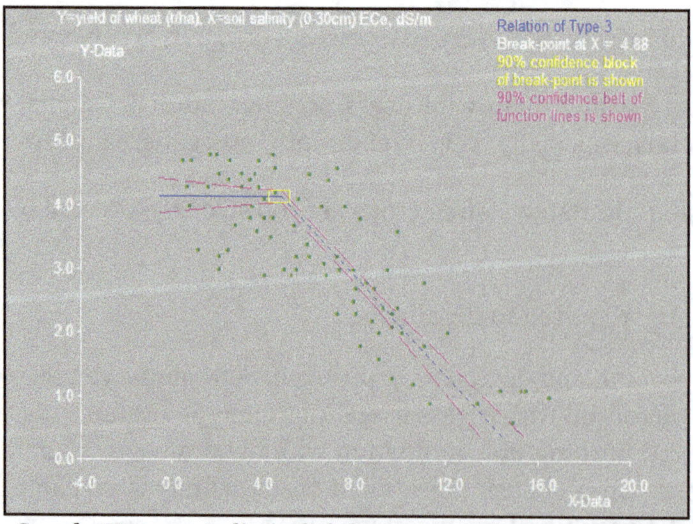

Wheat grown in Sampla, Haryana, India, is slightly sensitive, tolerating an ECe value of 4.9 dS/m.

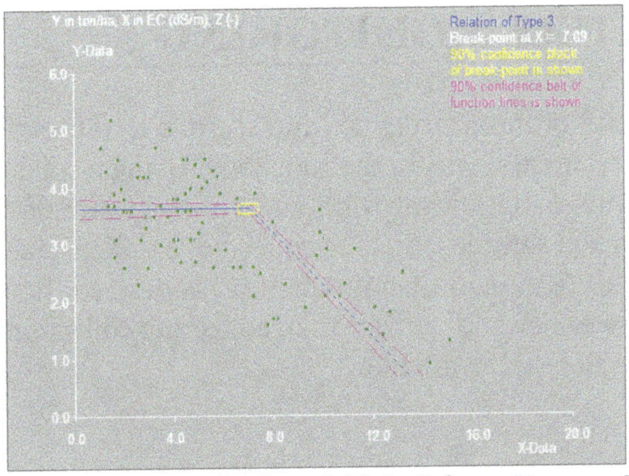

The field measurements in wheat fields in Gohana, Haryana, India, showed a higher tolerance level of ECe = 7.1 dS/m.

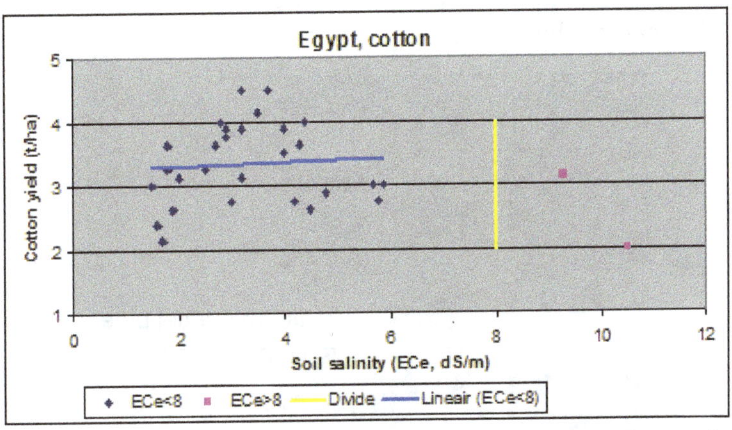

The cotton grown in the Nile Delta can be called salt-tolerant, with a critical ECe value of 8.0 dS/m. However, due to scarcity of data beyond 8 dS/m, the maximum tolerance level cannot be precisely determined and may actually be higher than that.

Regions Affected

From the FAO/UNESCO Soil Map of the World the following salinised areas can be derived.

Region	Area (10^6 ha)
Africa	69.5
Near and Middle East	53.1
Asia and Far East	19.5
Latin America	59.4
Australia	84.7
North America	16.0
Europe	20.7

SOIL MOISTURE MONITORING

The most basic method of monitoring soil moisture is to take soil samples at various depths (e.g. 1-foot increments) within the root zone. Using a soil auger, sampling tube, or even a shovel requires significant time and effort to sample throughout the root zone. An undisturbed soil samples must be collected, a difficult task if a volumetric soil moisture measurement (e.g. in/ft of soil moisture) is desired. In order to make monitoring of soil moisture easier and quicker than collecting soil samples, a number of soil moisture monitoring tools are available.

Tensiometers

A tensiometer is a device for measuring soil water tension. It consists of a cylindrical pipe about one inch in diameter with a porous ceramic cup attached to one end and vacuum gauge attached to the other.

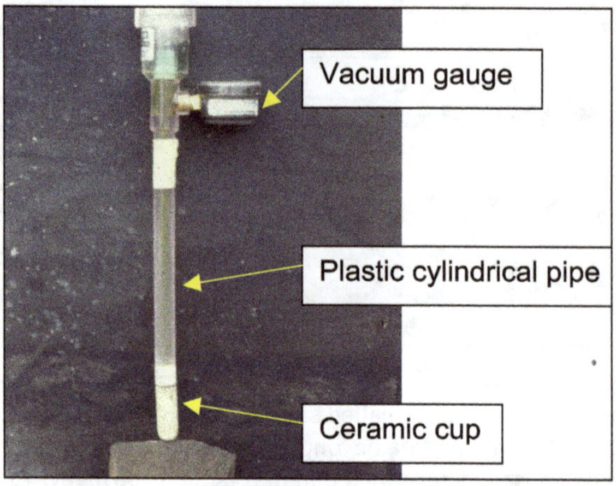

Tensiometer.

Tensiometers indirectly measure soil moisture tension. Since tensiometers are installed for the entire season or longer, they provide readings at the same location over an extended period of time. Tensiometer readings are easily interpreted and indicate the soil water conditions experienced by the trees' roots. Soil salinity does not affect the readings.

Although tensiometers are used most frequently for monitoring soil moisture, they can also be incorporated into automated irrigation systems. Tensiometers with solenoids can be used to control an irrigation system and tensiometers equipped with transducers can be used with computerized irrigation systems.

Installation

To measure soil water tension, the end of the tensiometer having the porous cup is inserted at the depth desired through a pilot hole, made with a soil probe, in the soil. The porous cup should be soaked in water for several hours before installation. After installation, the tensiometer is filled with water and allowed to equilibrate with the soil water for about twenty-four hours. A tensiometer should be installed in the zone of greatest root density, at about one-quarter to one-third of the maximum root depth. A tensiometer at this depth can be used to schedule irrigations. A tensiometer reading of Ceramic cup Vacuum gauge Plastic cylindrical pipe approximately 60 centibars indicates a need for irrigation in the near future. Refinements based on soil sampling will be required to adjust for site-specific conditions.

It is recommended that a second tensiometer also be installed in the bottom onethird of the root depth, to assure that the moisture extends to an adequate depth. Frequently, the shallower tensiometer will begin to show an increased reading (indicating drier soil) prior to the deeper tensiometer beginning to change. Trees tend to take up water from the shallower depths first. Some water managers utilizing tensiometers use the beginning of moisture extraction from the deeper depths, indicated by a change in the lower tensiometer's readings, as an indicator of when to irrigate. This practice seems to work well as long as irrigation water can penetrate to the lower depths to replace the soil moisture used. If the tensiometer reading at the lower root depth remains unchanged following an irrigation or continues to rise during the growing season, irrigation applications may be insufficient.

The number of tensiometer stations required depends on the irrigation system and on soil uniformity and management. For areas up to forty acres, at least two stations should be established. Stations should be located in areas representative of overall moisture status, with separate stations for problem areas or for areas having different soil conditions.

What do the Readings Mean?

In an unsaturated soil, soil water tension—frequently called the "suction" — falls below atmospheric pressure. As wet soil dries, the soil-water suction increases, causing water to flow out of the tensiometer through the porous cup. The small pores of the saturated cup prevent air from entering the tensiometer. This outflow of water creates a vacuum inside the tensiometer and increases the reading on the vacuum gauge. If the soil is re-wetted by irrigation, water will be drawn back into the tensiometer, reducing the vacuum inside, and the reading on the tensiometer gauge will decrease. The vacuum gauge measures the suction in centibars (100 centibars = 1 Bar = 1 Atm), with a range of 0 to 100. A reading of zero indicates a saturated soil in which plant roots will probably suffer from poor aeration. A reading of 10 to 25 centibars reflects a soil at field capacity. The lower reading is for sandy soils at field capacity, and the higher reading is for finertextured soils. Readings of 70 to 80 indicate a dry soil. Tensiometers will not read above 85 centibars.

Tensiometers do not provide information on the amount of water depleted from the soil unless they have been calibrated for the particular soil type. They therefore indicate when to irrigate, but not how much to irrigate.

Maintenance

Tensiometers must be properly maintained. This requires periodically filling the pipe with water and replacing porous cups. If the soil becomes too dry (tensiometer readings greater than 85 centibars), the porous cup will break tension and air will enter the tensiometer. A cracked cup will also prevent a vacuum from developing in the tensiometer and cause the instrument to always read zero. In locations where temperatures fall below freezing, the tensiometers should be protected or removed from the field. Finally, the porous cup of a tensiometer filled with water should not be exposed to the atmosphere for long periods of time. Such exposure causes evaporation of water from the cup's surface, which in turn causes salt buildup and clogging of the cup.

Electrical Resistance Blocks

Electrical resistance blocks, sometimes referred to as gypsum blocks, are an inexpensive and simple soil moisture measurement tool. They are useful for timing irrigations, but do not provide information regarding the amount of irrigation water required. Electrical resistance blocks are installed permanently at a site and have the advantages of allowing their user to quickly monitor soil moisture at the same location throughout the season. Gypsum blocks monitor soil moisture indirectly by measuring the electrical resistance between two electrodes attached to a small, cast block of gypsum buried in the soil. Some electrical resistance blocks have electrodes mounted in fiberglass or in perforated metal-encased blocks containing a sand/gypsum mixture. Wires, available in various lengths, lead from the blocks to the soil surface. The electrical resistance is read with a portable conductance meter. Conductance (1/resistance) is actually what is measured using the gypsum blocks. Measuring conductance requires use of alternating current (AC), which the meter produces from its direct current (DC) battery. Use of a simple resistance or ohm meter to read the electrical resistance blocks will not give stable or reliable readings. Electrical resistance blocks and meters vary in cost by manufacturer with blocks ranging in cost from $6-$35 each and meters costing approximately $250. Since the meter is portable, only a single meter is required to monitor all the installation sites. At a normal site, a gypsum block may last 2 to 3 years, but in areas of frequent irrigation or high water table it may require annual replacement.

One manufacturer produces a resistance block and meter which reads directly in centibars of soil water tension while most other manufacturers provide charts or tables to correlate the meter readings to soil water tension values. Soil water tension information indicates the timing of irrigations but does not tell the amount of irrigation water required. Indirectly, the irrigation amount can be arrived at through a trial-and-error procedure of monitoring the resistance blocks before and after irrigations. Blocks whose readings indicate wetter conditions after an irrigation reflect irrigation water reaching their depth. If a block does not reflect wetter soil conditions following an irrigation, it indicates that the irrigation was insufficient to reach the block's depth. Following the changes in resistance block readings through a number of irrigations will allow an irrigator to gauge the irrigation to match the soil moisture deficit.

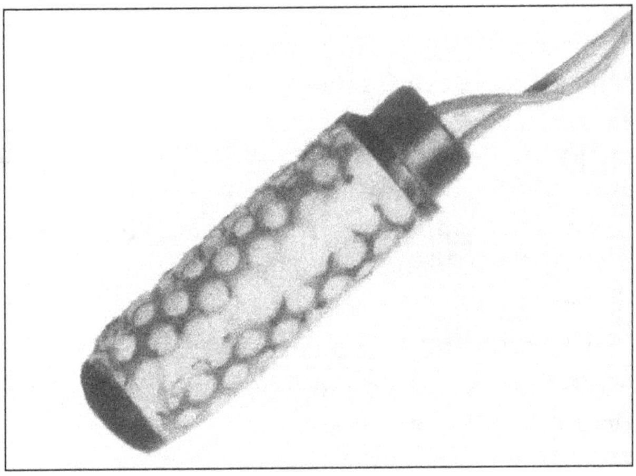

Electrical resistance block.

Data logging devices are also available which keep track of the readings of multiple resistance blocks connected to the logger. This has the advantage of keeping a "continuous" record of the soil moisture.

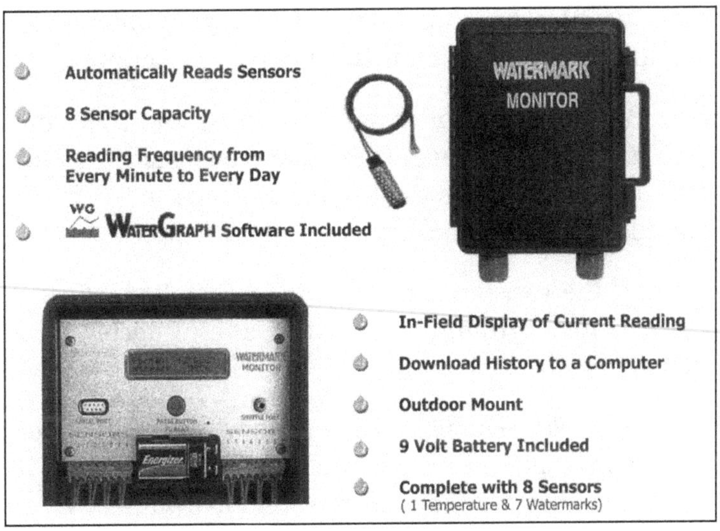

Installation

Electrical resistance blocks should be placed at several soil depths. At a minimum, blocks should be placed in the top one-third and bottom one-third of the root zone. Consultants who install soil moisture monitoring sites utilizing electric resistance blocks often place blocks at 1-foot intervals throughout the root zone. Soil moisture conditions are only monitored in the soil closely surrounding a resistance block.

Wires from the blocks are brought to the soil surface to facilitate periodic connection to the portable meter, or permanent connection to the data logger. The matric potential of the blocks is assumed to be in equilibrium with the surrounding soil, so the blocks act much like the surrounding soil - taking up and releasing water as the soil wets and dries. The electrical conductance between the electrodes varies according to the water content of the block. The higher the water content, the higher the conductance and the lower the electrical resistance.

The placement of resistance block stations in an field depends primarily on the soils. Generally, monitoring stations are chosen to be representative of what is happening in the surrounding area. At a minimum, two stations per 40 acres is recommended with additional stations being added in problem areas or in areas with different soil conditions. Since gypsum is soluble, blocks slowly dissolve. The life of a block can be extended by 1 or 2 years by the addition of a small quantity of gypsum to the back-fill soil at the time of installation. A small quantity of lime ($CaCO_3$) may likewise be useful in a acid soil in prolonging the life of a block.

Most electrical resistance blocks monitor soil moisture more effectively in the drier range of soil water tensions (in excess of 0.33 bar). Consequently, blocks are more useful in the medium to heavy textured soils (loams and clays), which retain more available water in the higher soil water tension range. Sands and coarse-textured soils tend to release much of their water at low soil water tensions, where the blocks are not as accurate nor responsive. Blocks are also affected by highly saline-alkali soil conditions and may therefore not be appropriate for use under such conditions.

In trees and vines, the blocks can often be installed in the tree/vine row to protect them from traffic or cultivation damage. They should be installed within the drip line of the tree. When sprinkler irrigation is used, avoid placing the tensiometers in the drier "shadow" area caused by the tree blocking the sprinkler's spray.

In field and row crops, the resistance blocks can be installed where desired with the wire leads being placed underground, terminating at a protected location where they can be accessed for monitoring.

At a station, resistance blocks to be placed at various depths can either be sequentially stacked in the same hole or placed in separate holes. Placing blocks in separate holes has the advantage of locating the block in the soil environment closest to an undisturbed condition, but the disadvantage of requiring more work than installing multiple blocks in a single hole. Resistance block users have used both techniques successfully.

Prior to installation, each block should be tested by soaking it in water and hooking it up to its companion conductance meter. The installation hole should be only slightly larger than the block and can be made using a soil probe or small auger. The hole should be about 1 inch deeper than the depth at which you want to install the block. Before inserting the block, put a small quantity of loose, moist soil (to which you can add small quantities of gypsum and lime) into the hole. About 2 ounces of water should be added to the hole to further moisten the loose soil in the bottom of the hole.

To place the block in the hole, run the wire lead through an appropriate length of 1/2-inch diameter PVC pipe and put tension on the wire to hold the block on the end of the pipe. One manufacturer's block has a special collar which fits into the end of 1/2-inch PVC pipe to facilitate installation. Lower the block into the hole and push it firmly into the loose, moist soil in the bottom of the hole. Remove the PVC pipe and fill the hole with the soil removed in making the hole or with a soil-gypsum-lime mixture for several inches. Pack the soil firmly, preferably using a wooden rod such as a broom handle so as to avoid damaging the blocks or wires. Continue refilling the hole in stages making sure that the soil is firmly packed at each stage in an effort to simulate undisturbed conditions. Refilling the hole properly is important to ensure that water and roots do not penetrate the hole more easily than the surrounding soil.

When installing blocks in traffic or cultivated areas, you can trench or bury the lead wires to prevent damage. Identify the blocks by knotting the wire, color coding, or by tagging. If possible, protect the bare ends of the wire leads since they tend to become corroded. Scraping the wire leads with a knife blade or fingernail file prior to attaching the meter will facilitate a better connection.

How to use the Information from the Blocks

Electric resistance block information is best used by evaluating changes in block readings. A block which shows that the surrounding soil is becoming drier between irrigations, is a good indicator that water uptake by roots is occurring at that depth. Most frequently, plants preferentially take up water from the shallower depths; thus drying the shallow depths down prior to taking up water from deeper depths. The shallower-placed block therefore shows a change first. Some block users utilize this transition of water uptake from the drier, shallow depths to the deeper depths, along with a particular meter reading of the shallow block, as a signal to irrigate. These "signals" to irrigate are often arrived at over time by watching and comparing the tree response and block readings.

As mentioned earlier, the blocks do not provide information on how much to irrigate, only on when to irrigate. By trial-and-error, irrigation amounts can be estimated. Reading the blocks just prior to irrigation and then reading them 2-3 days after irrigation (drainage has nearly stopped) can provide information on the depths to which irrigation water penetrated. If a blocks reading's remained unchanged (assuming it is not already reading completely wet), indicates that water did not reach that depth. Assuming you wanted to wet to that depth, more water - a longer irrigation - would need to be applied.

If the block(s) at the deeper depths never change readings, this indicates that you may be able to go longer between irrigations or that there is no root activity at that soil depth. A lack of root activity can be due to saturated soil conditions (resulting from over-irrigation and/or a high water table), disease, or nematode problems. In addition, if the deeper-depth blocks continually read wet, you cannot tell if you are over-irrigating and losing water to deep percolation.

If there is a question as to the accuracy of a block(s) readings, often indicated by a block's readings remaining unchanged when you think it should be changing, checking a nearby location's soil moisture using a soil auger and the "feel" method is an excellent first check.

Neutron Moisture Meters

Neutron soil moisture meters (neutron probe) use radioactive material for measuring soil moisture. They contain an electronic gauge, a connecting cable, and a source tube containing both nuclear source and detector tube. An access tube is installed in the ground, and the source tube is lowered into the tube to the desired depths of measurement.

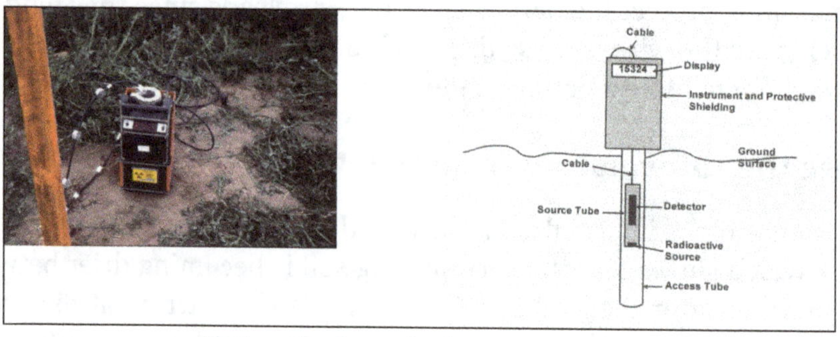

Photo and schematic of a neutron probe.

An Americium 241/Beryllium pellet emits high-energy neutrons (typically with a radiation rating of 50 millicurries/m). Lowering the source tube down in soil causes high-speed neutrons to collide with hydrogen atoms in water and soil. Their energy is then lost and lowenergy or "slow" neutrons are created. Some slow neutrons are reflected back to the source tube and counted by the neutron detector. These raw counts are transmitted to the gauge that contains a microprocessor with a calibration equation (supplied by the user or manufacturer) to give a reading of volumetric soil moisture content. Because soil moisture is the primary source of hydrogen atoms in the soil, neutron moisture meters indirectly measure soil moisture content.

An advantage of these meters is a volume of soil about the size of a volleyball is sampled. This better reflects soil moisture contents affecting plant growth compared with the small sampling volumes of other moisture sensors. Also, measurements at multiple depths can be made at a given site in a few minutes.

One disadvantage of this device is that reliable measurements at shallow depths (less than six inches) may not be possible because some neutrons escape from the soil surface into the air instead of being detected. The main disadvantage of this method, however, is its operator training, storage, and licensing and inspection requirements because the radioactive source.

Installing a Neutron Moisture Meter

Select a site that will not be disturbed by equipment. Install an access tube in the soil to a depth of about 6 inches deeper than the maximum measurement depth. A small gap between access tube and soil can be tolerated, but surface water should not flow down the side of the access tube. Cover the top of the access tube between measurements to keep water, frogs, and other foreign material out of the tube.

Two-inch diameter aluminum pipe, Class 125 PVC pipe, or electrical metal tubing (EMT) is commonly used for access tubes. Because meter readings are affected by access tubing material, the same material must be used for each site. Aluminum affects the readings the least, while PVC affects the readings the most because "slow neutrons" are absorbed by chlorine in PVC. In high water table areas, the bottom of the access tubing must be sealed to keep out water. Water in the access tube can damage the source tube and cause very high readings.

What do the Readings Mean?

Actual counts are not very useful. Actual counts must be related to volumetric soil moisture contents expressed in inches of water per foot of soil using an appropriate calibration curve. Then, measured soil moisture contents can be compared with soil moisture contents at field capacity, and soil moisture depletion determined. Actual depletion can be compared with allowable depletion to determine when to irrigate and how much to apply.

Multiple calibration curves can be stored in the memory of the electronics of the meter and can provide direct readouts of soil moisture content in a variety of terms, such as inches per foot and soil moisture percentage. Field readings can be stored for later retrieval or transfer to computers.

Placement of Monitoring Sites

At least two sites for each 40 acres are recommended. More sites may be necessary depending on soil variability. Additional sites may be needed for problem areas or for areas with different soil or management conditions.

The maximum depth of access tubes should be at least the depth of the root zone. However, measuring moisture contents at deep depths may be desirable for monitoring percolation below the root zone.

Maintenance

Neutron meters require little maintenance beyond checking to ensure proper operation. Access tubes should be checked for water or foreign materials in them. The most common failure is a broken or worn cable, which connects the source tube to the electronic readout device. A repaired instrument may also require recalibration. Operator recertification is required every two years.

Considerations in using Neutron Moisture Meters

- Access Tubes: Access tubing material can affect counting rates. Highest counting rates are achieved with aluminum, lowest in PVC. However, accurate calibration curves can be made with PVC, which is less expensive than aluminum and does not easily bend or become misshapen.

- Standard Counts: A standard count is made periodically to monitor performance of the meter and to calculate count ratios (actual count divided by standard count). Standard counts should be taken with the detector/source tube locked in the polypropylene shielding positioned on top of the transport case. The instrument should be at least 2 feet from any material that can influence the count. At least five counts are recommended for calculating the average standard count to minimize errors in the count ratio.

- Air Gaps: A slight air gap between access tube and soil will not greatly affect count rate. However, prevent water from flowing down the gap. Root density also might be affected by gaps. Random pockets of clay in an otherwise sandy loam root zone of an orchard will give artificially high readings of water content that do not reflect the average of the root zone. Avoid these areas.

- Count Accuracy: Theoretically, one reading per depth is sufficient for an error of one percent or less if 5,000 or more slow neutrons are counted. Most neutron moisture meters base their standard count on a 16 second duration.

- Count Duration: A 15 or 16-second count duration is a common count time for most soils. A longer count time may be needed in sandy soil. However, a 4-second count can also be used with reasonable accuracy when measurements are to be made at many sites and depths. The standard error of 0.1 inch of water/foot of soil was found for the 4-second count and 0.06 inch/ft for the 16-second count.

- Interchanging Meters: Use caution in interchanging neutron moisture meters. Meters with similar standard counts and their calibration curves can be interchanged as long as count ratios are used to standardize the meter's performance. Meters with very different standard counts should not be interchanged unless calibration curves for each instrument are available. Using standard counts may not standardize performances of very different meters.

- Depth of First Reading: Counts made at a six inch depth below the surface are affected by the soil-air interface, especially as the soil dries after irrigation, due to the large sampling volume of the meter. Thus, a separate calibration curve may be needed for accurate measurements made at that depth. In practice, however, making the first reading at 9 inches deep can minimize the effect of the soil-air interface.

Dielectric Soil Moisture Sensors

Dielectric soil moisture sensors determine the soil moisture by measuring the dielectric constant of the soil, an electrical property that is highly dependent on the moisture content. The constant for a dry soil is between 3 and 5; about one for air; and is about 80 for water. Thus, changes in the moisture content cause a substantial change in the dielectric constant of the soil. Calibration equations correlate the dielectric constant with soil moisture content. The most common dielectric devices are capacitance sensors and time-domain-reflectometry (TDR) sensors although other types of dielectric sensors exist. Virtually all of the sensors listed below can be attached to data loggers for a continuous record of soil moisture content. Many of these systems have the option of radio or cell phone telemetry that can link the sensors directly to an office computer or be viewed through a web-based server.

Capacitance Sensors

The capacitance sensor consists of two electrodes separated by a material called the dielectric, a material that does not readily conduct an electrical current. Normally, cylindrical-shaped electrodes are used. Inserting the sensor into the soil results in the soil becoming part of the dielectric. An oscillator applies a frequency between 50 and 150 Mhz (depending on the manufacturer) to the electrodes, which, in turn, results in a resonant frequency, the magnitude of which depends on the dielectric constant of the soil. The greater the soil moisture content, the smaller the frequency. The frequency is then used with a calibration equation to estimate volumetric soil moisture content.

Capacitance instrument.

Time-domain-reflectometry (TDR)

Time-domain-reflectometry involves installing two or three steel rods, called waveguides, into the soil parallel to each other. Components of a TDR system are a voltage pulse generator, a signal analyzer, the waveguides, and a cable connecting the waveguides to the instrumentation. An electrical pulse applied to the waveguides travels their length, and then is reflected back when it reaches the end of the waveguides. The travel time required for the pulse to reach the end of the waveguides and back depends on the dielectric constant of the soil. The larger the dielectric constant, the longer the pulse travel time.

TDR probe.

Calibration

Calibration equations relating the dielectric constant to the soil moisture content are necessary. These equations generally are provided by the manufacturer; however, in some cases, site-specific calibration may be needed. Some manufacturers provide a universal calibration, while others provide calibrations based on soil texture. Some calibration curves provide the volumetric soil moisture content. Others may display the percent of available soil moisture remaining in the soil.

Field calibration may be necessary in fine-textured soils of California even though a manufacturer calibration curve is provided. Field experience has shown that unrealistically high readings can occur in these soils. If the sensor reading just after an irrigation (when the soil is at field capacity) exceeds 50% (6 inches/foot), then the sensor should be calibrated for the site-specific conditions. To determine if the dielectric sensor reading is realistic, compare the field capacity volumetric soil moisture content with the sensor value. If the values differ greatly, then a site-specific calibration is needed. Also, in some cases, very low sensor readings (approaching zero) have been found just before irrigation, indicating a need for calibration.

Developing a calibration curve involves collecting soil samples very close to the sensor and determining their volumetric soil moisture content and making a sensor reading at that depth at the same time.

Installation

All dielectric soil moisture sensors must be carefully installed to prevent air gaps between sensor and soil. Air gaps cause errors in their readings. Some sensors require specialized equipment for installation, while others are installed by placing the sensor into a pilot hole. In some cases, slurry is used to obtain a tight seal between sensor and soil.

Zone of Influence

The zone of influence is very small compared to the neutron moisture meter. The zone of influence is about 4 inches from the sensor for capacitance instruments and about 1 inch for TDR sensors.

Advantages/Disadvantages

Advantages of dielectric soil moisture sensors include the ability to be left in place to continuously log soil moisture content; repeatability of measurements; sensitivity to small changes in soil moisture content; and their precise resolution with depth because of the narrow vertical zone of influence. Disadvantages include the need for a calibration equation; the difficulty in developing the equation; the relatively small zone of influence; possible influence of soil salinity on probe reading; and sensitivity to air gaps.

Types of Instruments

Many types of dielectic soil moisture sensors exist. An technical report, Soil Water Monitoring: An Information Package, written by P. Charlesworth (CSIRO/CRC Irrigation Futures, Australia), which describes many of the dielectric sensors used world-wide.

Field Studies

Several studies have compared the readings of some dielectric soil moisture sensor with the soil moisture content measured with a neutron moisture meter, calibrated for a specific site. Default calibration curves of the dielectric instruments were used.

Results showed that most of the dielectric sensors generally followed the trends in soil moisture content; however, one instrument tracked moisture content changes very poorly. In general, these sensors perform best in the coarser-textured, non-saline soils. These results also showed that the readings of the dielectric sensors can differ considerably from those of calibrated neutron moisture meter data, indicating the need for soil specific calibration.

References

- Soil: newworldencyclopedia.org, Retrieved 5 February, 2020
- Environmental protection agency (ireland) (2017). National ambient air quality monitoring programme 2017–2022. Environmental protection agency (ireland). P. 30. Isbn 9781840957501. Retrieved 31 may 2018
- Receptor modeling". Air quality management online portal. U.s. Environmental protection agency. Archived from the original on 3 september 2014. Retrieved 31 may 2018
- Panagos, p., van liedekerke, m., yigini, y., montanarella, l. (2013) contaminated sites in europe: review of the current situation based on data collected through a european network. Journal of environmental and public health in press. Doi:10.1155/2013/158764
- Agarwal, a.; liu, y. (2015). "remediation technologies for oil-contaminated sediments". Marine pollution bulletin. Doi:10.1016/j.marpolbul.2015.09.010
- Soil-pollutant, earth-and-planetary-sciences: sciencedirect.com, Retrieved 5 February, 2020
- "free articles and software on drainage of waterlogged land and soil salinity control". Retrieved 2010-07-28

5

Water Quality Monitoring

The chemical, biological and physical, and radiological characteristics of water can be termed as water quality. Its monitoring refers to the assessment of the qualitative information of these characteristics of water. This chapter closely examines these key concepts of water quality monitoring to provide an extensive understanding of the subject.

WATER QUALITY

Water quality is measured by several factors, such as the concentration of dissolved oxygen, bacteria levels, the amount of salt (or salinity), or the amount of material suspended in the water (turbidity). In some bodies of water, the concentration of microscopic algae and quantities of pesticides, herbicides, heavy metals, and other contaminants may also be measured to determine water quality.

Although scientific measurements are used to define water quality, it is not a simple thing to say "that water is good" or "that water is bad." So, the determination is typically made relative to the purpose of the water – is it for drinking or to wash a car with or for some other purpose?

Poor water quality can pose a health risk for people. Poor water quality can also pose a health risk for ecosystems.

Water quality is affected by a wide range of natural and human influences. The most important of the natural influences are geological, hydrological and climatic, since these affect the quantity and the quality of water available. Their influence is generally greatest when available water quantities are low and maximum use must be made of the limited resource; for example, high salinity is a frequent problem in arid and coastal areas. If the financial and technical resources are available, seawater or saline groundwater can be desalinated but in many circumstances this is not feasible. Thus, although water may be available in adequate quantities, its unsuitable quality limits the uses that can be made of it. Although the natural ecosystem is in harmony with natural water quality, any significant changes to water quality will usually be disruptive to the ecosystem.

The effects of human activities on water quality are both widespread and varied in the degree to which they disrupt the ecosystem and/or restrict water use. Pollution of water by human faeces, for example, is attributable to only one source, but the reasons for this type of pollution, its impacts on water quality and the necessary remedial or preventive measures are varied. Faecal pollution may occur because there are no community facilities for waste disposal, because collection and treatment facilities are inadequate or improperly operated, or because on-site sanitation facilities (such as latrines) drain directly into aquifers. The effects of faecal pollution vary. In developing countries intestinal disease is the main problem, while organic load and eutrophication may be of greater concern in developed countries (in the rivers into which the sewage or effluent is discharged and in the sea into which the rivers flow or sewage sludge is dumped). A single influence may, therefore, give rise to a number of water quality problems, just as a problem may have a number of contributing influences. Eutrophication results not only from point sources, such as wastewater discharges with high nutrient loads (principally nitrogen and phosphorus), but also from diffuse sources such as run-off from livestock feedlots or agricultural land fertilised with organic and inorganic fertilisers. Pollution from diffuse sources, such as agricultural runoff, or from numerous small inputs over a wide area, such as faecal pollution from unsewered settlements, is particularly difficult to control.

The quality of water may be described in terms of the concentration and state (dissolved or particulate) of some or all of the organic and inorganic material present in the water, together with certain physical characteristics of the water. It is determined by in situ measurements and by examination of water samples on site or in the laboratory. The main elements of water quality monitoring are, therefore, on-site measurements, the collection and analysis of water samples, the study and evaluation of the analytical results, and the reporting of the findings. The results of analyses performed on a single water sample are only valid for the particular location and time at which that sample was taken. One purpose of a monitoring programme is, therefore, to gather sufficient data (by means of regular or intensive sampling and analysis) to assess spatial and/or temporal variations in water quality.

The quality of the aquatic environment is a broader issue which can be described in terms of:

- Water quality.
- The composition and state of the biological life present in the water body.
- The nature of the particulate matter present.
- The physical description of the water body (hydrology, dimensions, nature of lake bottom or river bed, etc).

Complete assessment of the quality of the aquatic environment, therefore, requires that water quality, biological life, particulate matter and the physical characteristics of the water body be investigated and evaluated. This can be achieved through:

- Chemical analyses of water, particulate matter and aquatic organisms (such as planktonic algae and selected parts of organisms such as fish muscle).
- Biological tests, such as toxicity tests and measurements of enzyme activities.
- Descriptions of aquatic organisms, including their occurrence, density, biomass, physiology and diversity (from which, for example, a biotic index may be developed or microbiological characteristics determined).
- Physical measurements of water temperature, pH, conductivity, light penetration, particle size of suspended and deposited material, dimensions of the water body, flow velocity, hydrological balance, etc.

Pollution of the aquatic environment, as defined by GESAMP, occurs when humans introduce, either by direct discharge to water or indirectly (for example through atmospheric pollution or water management practices), substances or energy that result in deleterious effects such as:

- Hazards to human health.
- Harm to living resources.
- Hindrance to aquatic activities such as fishing.
- Impairment of water quality with respect to its use in agriculture, industry or other economic activities, or reduction of amenity value.

The importance attached to quality will depend on the actual and planned use or uses of the water (e.g. water that is to be used for drinking should not contain any chemicals or microorganisms that could be hazardous to health).

Since there is a wide range of natural water qualities, there is no universal standard against which a set of analyses can be compared. If the natural, pre-polluted quality of a water body is unknown, it may be possible to establish some reference values by surveys and monitoring of unpolluted water in which natural conditions are similar to those of the water body being studied.

Characteristics of Surface Waters

Hydrological Characteristics

Continental water bodies are of various types including flowing water, lakes, reservoirs and groundwaters. All are inter-connected by the hydrological cycle with many intermediate water bodies, both natural and artificial. Wetlands, such as floodplains, marshes and alluvial aquifers, have characteristics that are hydrologically intermediate between those of rivers, lakes and groundwaters. Wetlands and marshes are of special biological importance.

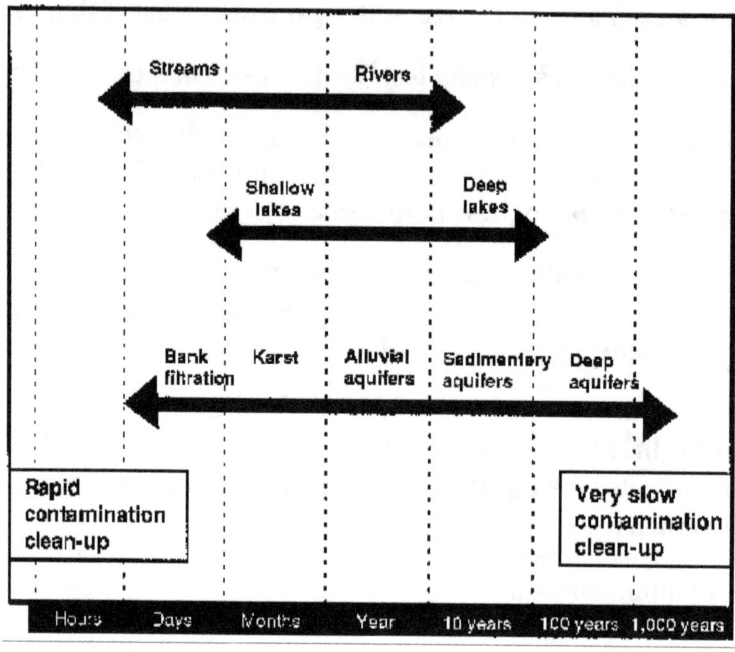

Typical water residence times in inland water bodies.

Actual residence times may vary. Residence times in karstic aquifers may vary from days to thousands of years, depending on extent and recharge. Some karstic aquifers of the Arabian peninsula have water more than 10,000 years old.

It is essential that all available hydrological data are included in a water quality assessment because water quality is profoundly affected by the hydrology of a water body. The minimum information required is the seasonal variation in river discharge, the thermal and mixing regimes of lakes, and the recharge regime and underground flow pattern of groundwaters.

The common ranges of water residence time for various types of water body are shown in figure. The theoretical residence time for a lake is the total volume of the lake divided by the total outflow rate ($V/\Sigma Q$). Residence time is an important concept for water pollution studies because it is associated with the time taken for recovery from a pollution incident. For example, a short residence time (as in a river) aids recovery of the aquatic system from a pollution input by rapid dispersion and transport of waterborne pollutants. Long residence times, such as occur in deep lakes and aquifers, often result in very slow recovery from a pollution input because transport of waterborne pollutants away from the source can take years or even decades. Pollutants stored in sediments take a long time to be removed from the aquatic system, even when the water residence time of the water body is short.

River flow is unidirectional, often with good lateral and vertical mixing, but may vary widely with meteorological and climatic conditions and drainage pattern. Still surface waters, such as deep lakes and reservoirs, are characterised by alternating periods of stratification and vertical mixing. In addition, water currents may be multi-directional and are much slower than in rivers. Moreover, wind has an important effect on the movement of the upper layers of lake and reservoir water. The residence time of water in lakes is often more than six months and may be as much as several hundred years. By contrast, residence times in reservoirs are usually less than one year.

Lakes and Reservoirs

An important factor influencing water quality in relatively still, deep waters, such as lakes and reservoirs, is stratification. Stratification occurs when the water in a lake or reservoir acts as two different bodies with different densities, one floating on the other. It is most commonly caused by temperature differences, leading to differences in density (water has maximum density at 4 °C), but occasionally by differences in solute concentrations. Water quality in the two bodies of water is also subject to different influences. Thus, for example, the surface layer receives more sunlight while the lower layer is physically separated from the atmosphere (which is a source of gases such as oxygen) and may be in contact with decomposing sediments which exert an oxygen demand. As a result of these influences it is common for the lower layer to have a significantly decreased oxygen concentration compared with the upper layer. When anoxic conditions occur in bottom sediments, various compounds may increase in interstitial waters (through dissolution or reduction) and diffuse from the sediments into the lower water layer. Substances produced in this way include ammonia, nitrate, phosphate, sulphide, silicate, iron and manganese compounds.

Temperate Lakes

Thermal stratification has been studied for many years in temperate regions where, during spring and summer, the surface layers of the water become warmer and their density decreases. They float upon the colder and denser layer below and there is a resistance to vertical mixing. The warm surface layer is known as the epilimnion and the colder water trapped beneath is the hypolimnion. The epilimnion can be mixed by wind and surface currents and its temperature varies little with depth. Between the layers is a shallow zone, called the metalimnion or the thermocline, where the temperature changes from that of the epilimnion to that of the hypolimnion. As the weather becomes cooler, the temperature of the surface layer falls and the density difference between the two layers is reduced sufficiently for the wind to induce vertical circulation and mixing in the lake water, resulting in an "overturn". This can occur quite quickly. The frequency of overturn and mixing depends principally on climate (temperature, insolation and wind) and the characteristics of the lake and its surroundings (depth and exposure to wind).

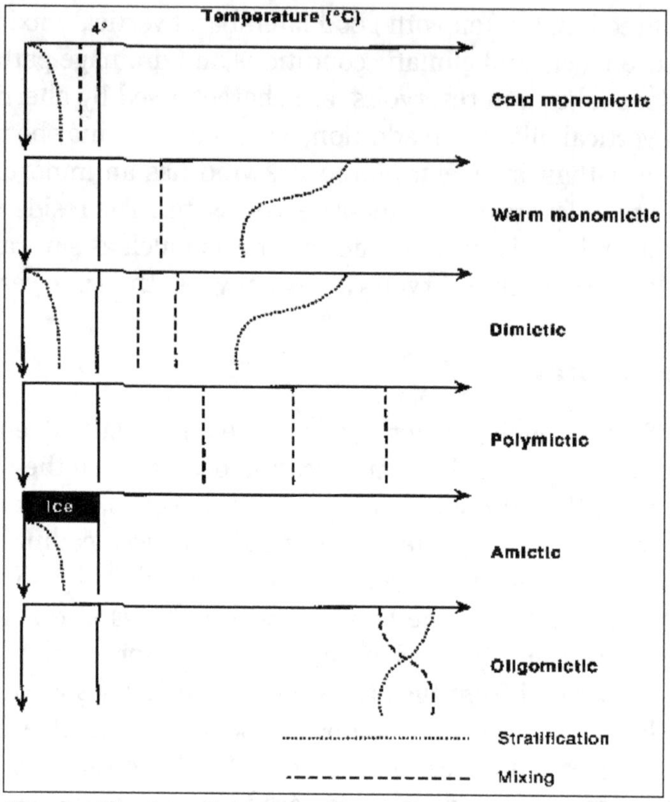

The classification of lakes according to the occurence of thermal stratification and mixing in the water column.

Lakes may be classified according to the frequency of overturn as follows:

- Monomictic: Once a year - temperate lakes that do not freeze.
- Dimictic: Twice a year - temperate lakes that do freeze.
- Polymictic: Several times a year - shallow, temperate or tropical lakes.
- Amictic: No mixing - arctic or high altitude lakes with permanent ice cover, and underground lakes.
- Oligomictic: Poor mixing - deep tropical lakes.
- Meromictic: Incomplete mixing - mainly oligomictic lakes but sometimes deep monomictic and dimictic lakes.

Thermal stratification does not usually occur in lakes less than about 10 m deep because wind across the lake surface and water flow through the lake tend to encourage mixing. Shallow tropical lakes may be mixed completely several times a year. In very deep lakes, however, stratification may persist all year, even in tropical and equatorial regions. This permanent stratification results in "meromixis", which is a natural and continuous anoxia of bottom waters.

Tropical Lakes

A common physical characteristic of tropical lakes is that seasonal variations in water temperature are small, as a result of relatively constant solar radiation. Water temperatures are generally high but decrease with increasing altitude. The annual water temperature range is only 2-3 °C at the surface and even less at depths greater than 30 m. Density differences are minimal because water temperature is almost constant. Winds and precipitation, both of which tend to be seasonal, play an important role in mixing. The very limited seasonal temperature variation also results in a correspondingly low annual heat budget in tropical lakes. However, the relative variation in the heat budget in consecutive years may be considerable, because the peak value of heat storage may result from a single meteorological event.

In some tropical lakes, variations in water level of several metres may result from the large differences in rainfall between wet and dry seasons. Such variations have pronounced effects on dilution and nutrient supply which, in turn, affect algal blooms, zooplankton reproduction and fish spawning.

During the dry season, wind velocities are generally higher than at other times of the year and evaporation rates are at their maximum. The resulting heat losses, together with turbulence caused by wind action, promote mixing.

The classification of lakes based on seasonal temperature variations at different depths is not generally applicable to tropical lakes. A classification which considers size, depth and other physical characteristics, such as the following, is more relevant:

- Large, deep lakes all have a seasonal thermocline in addition to a deep permanent thermocline over an anoxic water mass. Recirculation of the deep water may occur but the responsible mechanism is not clear.
- Large, shallow lakes have a distinct diurnal temperature variation. Temperature is uniform in the morning, stratification develops in the afternoon and is destroyed during the night. The fluctuation in water level may be considerable relative to lake volume and the large flood plain that results will have profound effects on the productivity of biological life in the water.
- Crater lakes generally have a small surface area relative to their great depth and are often stratified. Despite such lakes being in sheltered positions, special weather conditions can cause complete mixing of lake contents.
- High-altitude lakes in climates where there is only a small diurnal temperature difference are unstable and experience frequent overturns. Where temperature differences are larger, a more distinct pattern of stratification can be identified. There may also be substantial losses of water by evaporation during the night.
- River lakes are created when areas of land are flooded by rivers in spate. When

the water level in the river goes down, the lake water flows back towards the river. This annual or semiannual water exchange affects the biological and chemical quality of the water.

- Solar lakes. In saline, dark-bottomed lakes an anomalous stratification can develop. A lower, strongly saline water layer may be intensely heated by solar radiation, especially if it is well isolated from the atmosphere by the upper layer of lighter brine. Temperatures as high as 50 °C have been recorded in the lower levels of solar lakes.
- Temporary lakes occur in locations where the fluctuations of water level cause a shallow lake basin to dry up completely. In regions where there are pronounced wet and dry seasons this can occur annually, while in other regions the frequency of occurrence may be medium to long term. Temporary lakes often have an accumulation of salts on the lake bottom.

Rivers

An understanding of the discharge regime of a river is extremely important to the interpretation of water quality measurements, especially those including suspended sediment or intended to determine the flux of sediment or contaminants. The discharge of a river is related to the nature of its catchment, particularly the geological, geographical and climatological influences.

Tropical Rivers

The regime of a tropical river is largely determined by the annual cycle of wet and dry seasons. Some regimes, and some of the climatic and geographical conditions that affect regimes, are as follows:

- Equatorial rivers with one flow peak, resulting from heavy annual precipitation (1,750-2,500 mm) in areas with no marked dry season.
- Equatorial rivers with two flow peaks, produced by precipitation totalling more than 200 mm monthly and well over 1,750 mm annually. Equatorial forest predominates in the catchment area.
- Rivers in the moist savannah of tropical wet and dry lowlands exhibit pronounced seasonal effects of rainfall patterns. In these areas, the dry season persists for at least three months.
- In some areas of the tropical wet and dry highlands, the length of the dry season varies significantly. River basins in such areas are covered by woodland and relatively moist savannah.
- In the relatively drier regions of the tropical wet and dry highlands, river basins

are located in the marginal parts of the dry climate zones. Precipitation rarely exceeds 500-700 mm annually, which is typical of semi-desert regions, and vegetation in the river basins is predominantly dry savannah.

- In areas where the dry season is prolonged, ecologists may divide the associated vegetation into wooded steppe and grass steppe. River regimes, however, do not differ between the two types of region, and flow is likely to be intermittent.
- In desert regions, where annual rainfall is less than 200 mm, river basins are covered with sand, desert grass or shrubs, and rivers are of the wadi type. The drainage network is poor and where it is traceable, it is likely to have developed before the area reached its present stage of aridity.
- Many of the rivers of tropical mountain regions have drainage basins of very limited size.

The great rivers of the tropics do not fall exclusively into any of these categories, because their drainage basins extend over many regions of differing climate and vegetation. The regime of the Congo (Zaire) River, for example, is largely a combination of regimes of the equatorial wet region and the tropical highland climate. Mean monthly flow is highest in April (76,000 $m^3 s^{-1}$) and lowest in July (32,000 $m^3 s^{-1}$). The Niger has its headwaters in a wet zone near the ocean, but then flows into a semi-arid region where it is subject to evaporation losses. Lower still in its course, the river flows into wet and dry tropical lowland. The Zambesi river basin is located largely in wet and dry tropical highland and semi-arid regions, while the Nile exhibits the most complex regime of all African rivers, extending over many widely different climatic zones.

The Amazon has a complex flow regime because of the different precipitation patterns of its main tributaries. Flow begins to increase in November, reaches a peak in June and then falls to a minimum at the end of October. The Orinoco, although draining a similar area to the Amazon, reaches peak flow a month later than the Amazon but also has minimum flow in October.

Most of the large rivers of Asia that flow generally southward have their sources in the mountains and flow through varied climatic conditions before discharging to the sea. Peak flows generally occur when run-off from melting snow is supplemented by monsoon rains. The Ganges and Irrawady receive snow-melt from the Himalayas and southern Tibet respectively from April to June, and the flow rate is just beginning to decline when the July monsoon begins. Flooding can occur from July to October. The Mekong is somewhat similar. It has its beginnings at an altitude of about 4,900 m in China's Tanglha Range. Snow-melt is later here, so peak flows do not occur until August/September in the upper reaches of the river and October in the lower reaches. Minimum flow in the Mekong occurs from November to May.

In western Asia, the Indus has some of its source tributaries in the Hindu Kush mountains, while the Tigris and Euphrates rise in the mountains of Turkey and Armenia respectively. Monsoon rains cause the Indus to flood between July and September. The Tigris and Euphrates are affected by seasonal rains that overlap with snow-melt run-off and cause flooding from March to June. The floods on the Euphrates inundate low-lying areas to form permanent lakes that have no outlets. Water loss from these lakes is mainly by evaporation, although some of the water is withdrawn for irrigation.

Data on erosion in all of the world's river basins are far from complete. In general, however, erosion can be said to vary according to the following influences:

- Amount and pattern of rainfall and resultant river regime.
- Slope of the land.
- Extent of destruction of vegetation.
- Regeneration of vegetation.
- Soil type and resistance to the effects of temperature changes.

Table: Mean annual sediment loads of some major rivers.

River	Basin area (10^3 km^2)	Mean annual sediment load (10^6 t a^{-1})
Amazon	6,300	850
Brahmaputra	580	730
Congo (Zaire)	4,000	72
Danube	816	65
Ganges	975	1,450
Indus	950	435
Irrawady	430	300
Orinoco	950	150[1]
Mekong	795	160[1]
Ob	2,430	15
Rhine	160	2.8

Average discharge taken from Meybeck et al. taking into account existing dams.

Erosion rates are thus extremely variable, with highest rates occurring usually in mountain streams where human intervention has resulted in extensive damage to vegetation. Erosion is primarily responsible for the amount of sediment transported to the sea. The mean annual sediment loads transported by several major rivers are given in table.

Characteristics of Groundwater

Groundwater is held in the pore space of sediments such as sands or gravels or in the fissures of fractured rock such as crystalline rock and limestone. The body of rock or sediments containing the water is termed an aquifer and the upper water level in the saturated body is termed the water table. Typically, groundwaters have a steady flow pattern. Velocity is governed mainly by the porosity and permeability of the material through which the water flows, and is often up to several orders of magnitude less than that of surface waters. As a result mixing is poor.

The media (rock or sediment) in an aquifer are characterised by porosity and permeability. Porosity is the ratio of pore and fissure volume to the total volume of the media. It is measured as percentage voids and denotes the storage or water volume of the media. Permeability is a measure of the ease with which fluids in general may pass through the media under a potential gradient and indicates the relative rate of travel of water or fluids through media under given conditions. For water it is termed hydraulic conductivity.

Types of Aquifer

Underground formations are of three basic types: hard crystalline rocks, consolidated sedimentary formations and unconsolidated sediments. The hard crystalline rocks include granites, gneisses, schists and quartzites and certain types of volcanic rocks such as basalts and dolerites. These formations generally have little or no original porosity, and the existence of aquifers depends on fractures and fissures in the rock mass providing porosity and pathways for groundwater movement. Although these are often further enhanced by weathering, aquifers in hard rocks are usually small and localised and not very productive.

Groundwater in volcanic formations in regions of "recent" volcanic activity frequently contain fluoride and boron in concentrations that are unacceptably high for certain uses.

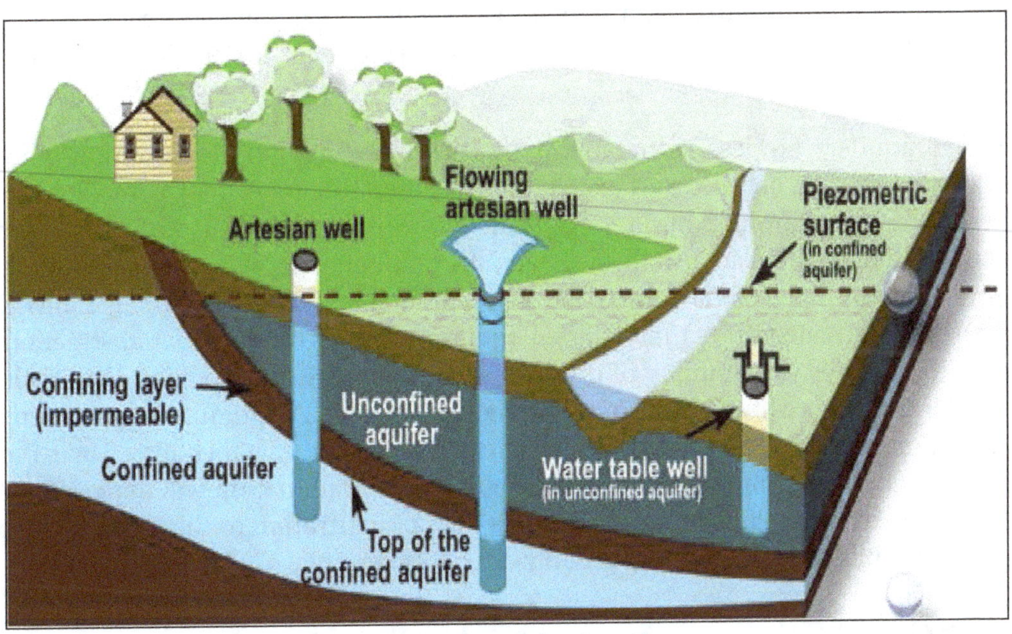

Confined and unconfined aquifers.

Consolidated sedimentary formations are often thick and extensive, and sometimes artesian. Limestone and sandstone formations may be highly porous and permeable and form some of the largest, most important and highest-yielding aquifers in the world. The permeability of these formations is largely due to fissures (fractures, faults, bedding planes). Porosity is also significant for the movement and storage of some pollutants. Dissolution of the rock can increase the permeability. The dissolution of carbonates, in particular, is responsible for the formation of karst aquifers, which can have large underground caverns and channels yielding substantial quantities of water.

Unconsolidated sediments occur as thin, superficial deposits over other rock types or as thick sequences in the major river or lake basins. Porosity and permeability are related to grain size. Sand and gravel deposits can provide important and high-yielding aquifers, whereas silts and clays are less productive. In the largest river basins, thick sedimentary deposits may contain many layers of different materials built up over long periods of time, producing important multi-aquifer sequences.

Aquifers may be confined or unconfined. A confined aquifer is overlain by an impermeable layer that prevents recharge (and contamination) by rainfall or surface water. Recharge of confined aquifers occurs where the permeable rock outcrops at or near the surface, which may be some distance from the area of exploitation. This feature may make control of quality and of pollution more difficult. Some aquifers are not perfectly confined and are termed semi-confined or leaky.

Unconfined aquifers are overlain by a permeable, unsaturated zone that allows surface water to percolate down to the water table. Consequently, they are generally recharged over a wide area and are often shallow with a tendency for interaction with surface water.

Confined aquifers are less vulnerable than unconfined aquifers to pollution outside their recharge zone because surface water and contaminants cannot percolate to the water table. If contamination does occur, however, it is often difficult to remedy because confined aquifers are usually deep and the number of points where contaminated water may be pumped out is limited. Given the limited outflow, contaminants may also be increasingly concentrated in confined aquifers and this may restrict abstraction of water. The greater vulnerability of unconfined aquifers to contamination is a result of the wider area over which they are recharged and in which contamination may enter, and the greater interaction with polluted surface water bodies which may lead to contaminant movement into groundwater. The risk of contamination will depend on the depth of the overlying unsaturated layer, the rate of infiltration to the water table and the land use in areas surrounding groundwater sources.

Water Quality

The quality of groundwater depends on the composition of the recharge water, the interactions between the water and the soil, soil-gas and rocks with which it comes into contact in the unsaturated zone, and the residence time and reactions that take place within the aquifer. Therefore, considerable variation can be found, even in the same general area, especially where rocks of different compositions and solubility occur. The principal processes influencing water quality in aquifers are physical (dispersion/dilution, filtration and gas movement), geochemical (complexation, acid-base reactions, oxidation-reduction, precipitation-solution, and adsorption-desorption) and biochemical (microbial respiration and decay, cell synthesis).

Groundwater quality is influenced by the effects of human activities which cause pollution at the land surface because most groundwater originates by recharge of rainwater infiltrating from the surface. The rainwater itself may also have an increased acidity due to human activity. The unsaturated zone can help reduce the concentrations of some pollutants entering groundwater (especially micro-organisms), but it can also act as a store for significant quantities of pollutants such as nitrates, which may eventually be released. Some contaminants enter groundwaters directly from abandoned wells, mines, quarries and buried sewerage pipes which by-pass the unsaturated zone (and, therefore, the possibility of some natural decontamination processes).

Contamination

Artificial pollution of groundwater may arise from either point or diffuse sources. Some of the more common sources include domestic sewage and latrines, municipal solid waste, agricultural wastes and manure, and industrial wastes (including tipping, direct injection, spillage and leakage). The contamination of groundwaters can be a complex process. Contaminants, such as agricultural chemicals, spread over large sections of the aquifer recharge area may take decades to appear in the groundwater and perhaps longer to disappear after their use has ceased. Major accidental spills and other point sources of pollutants may initially cause rapid local contamination, which then spreads through the aquifer. Pollutants that are fully soluble in water and of about the same density (such as chloride-contaminated water from sewage) will spread through the aquifer at a rate related to the groundwater flow velocity. Pollutants that are less dense than water will tend to accumulate at the water table and flow along the surface. Dense compounds such as chlorinated solvents will move vertically downwards and accumulate at the bottom of an aquifer.

There is usually a delay between a pollution incident and detection of the contaminant at the point of water abstraction because movement in the unsaturated zone and flow in the aquifer are often slow. For similar reasons the time needed to “flush out” a pollutant is long and in some cases the degradation of groundwater quality may be considered irreversible.

Land use in areas surrounding boreholes and where aquifers are recharged should be carefully monitored as part of a pollution control programme. The vulnerability of the aquifer to pollution will depend, in part, on the human activity and land use in areas where rainfall or surface water may percolate into the aquifer. In these areas, contamination of surface water or of the unsaturated layer above an aquifer is likely to cause groundwater pollution.

Natural Processes affecting Water Quality

Although degradation of water quality is almost invariably the result of human activities, certain natural phenomena can result in water quality falling below that required for particular purposes. Natural events such as torrential rainfall and hurricanes lead to excessive erosion and landslides, which in turn increase the content of suspended material in affected rivers and lakes. Seasonal overturn of the water in some lakes can bring water with little or no dissolved oxygen to the surface. Such natural events may be frequent or occasional. Permanent natural conditions in some areas may make water unfit for drinking or for specific uses, such as irrigation. Common examples of this are the salinisation of surface waters through evaporation in arid and semi-arid regions and the high salt content of some groundwaters under certain geological conditions. Many groundwaters are naturally high in carbonates (hardness), thus necessitating their treatment before use for certain industrial applications. Groundwaters in some regions contain specific ions (such as fluoride) and toxic elements (such as arsenic and selenium) in quantities that are harmful to health, while others contain elements or compounds that cause other types of problems (such as the staining of sanitary fixtures by iron and manganese).

The nature and concentration of chemical elements and compounds in a freshwater system are subject to change by various types of natural process, i.e. physical, chemical, hydrological and biological. The most important of these processes and the water bodies they affect are listed in table.

The effects on water quality of the processes listed in table will depend to a large extent on environmental factors brought about by climatic, geographical and geological conditions. The major environmental factors are:

- Distance from the ocean: Extent of sea spray rich in Na^+, Cl^-, Mg^{2+}, SO_4^{2-} and other ions.
- Climate and vegetation: Regulation of erosion and mineral weathering; concentration of dissolved material through evaporation and evapo-transpiration.
- Rock composition (lithology): The susceptibility of rocks to weathering ranges from 1 for granite to 12 for limestone; it is much greater for more highly soluble rocks (for example, 80 for rock salt).
- Terrestrial vegetation: The production of terrestrial plants and the way in which

plant tissue is decomposed in soil affect the amount of organic carbon and nitrogenous compounds found in water.

- Aquatic vegetation: Growth, death and decomposition of aquatic plants and algae will affect the concentration of nitrogenous and phosphorous nutrients, pH, carbonates, dissolved oxygen and other chemicals sensitive to oxidation/reduction conditions. Aquatic vegetation has a profound effect on the chemistry of lake water and a less pronounced, but possibly significant effect, on river water.

Table: Important processes affecting water quality.

Process type	Major process within water body	Water body
Hydrological	Dilution	All water bodies
	Evaporation	Surface waters
	Percolation and leaching	Groundwaters
	Suspension and settling	Surface waters
Physical	Gas exchange with atmosphere	Mostly rivers and lakes
	Volatilisation	Mostly rivers and lakes
	Adsorption/desorption	All water bodies
	Heating and cooling	Mostly rivers and lakes
	Diffusion	
Chemical	Photodegradation	
	Acid base reactions	All water bodies
	Redox reactions	All water bodies
	Dissolution of particles	All water bodies
	Precipitation of minerals	All water bodies
	Ionic exchange[1]	Groundwaters
Biological	Primary production	Surface waters
	Microbial die-off and growth	All water bodies
	Decomposition of organic matter	Mostly rivers and lakes
	Bioaccumulation[2]	Mostly rivers and lakes
	Biomagnification[3]	Mostly rivers and lakes

Ionic exchange is the substitution of cations, for example in clay; the commonest is the replacement of calcium by sodium.

Bioaccumulation results from various physiological processes by which an organism accumulates a given substance (mercury and lead are common examples) from water or suspended particles.

Biomagnification is the increase in concentration of a given substance within a food chain such as phytoplankton → zooplankton → microphage fish → carnivorous fish → carnivorous mammals.

Under the influence of these major environmental factors, the concentrations of many chemicals in river water are liable to change from season to season. In small watersheds (<100KM^2) the influence of a single factor can cause a variation of several orders of magnitude. Water quality is generally more constant in watersheds greater than 100,000 km^2, and the variation is usually within one order of magnitude for most of the measured variables.

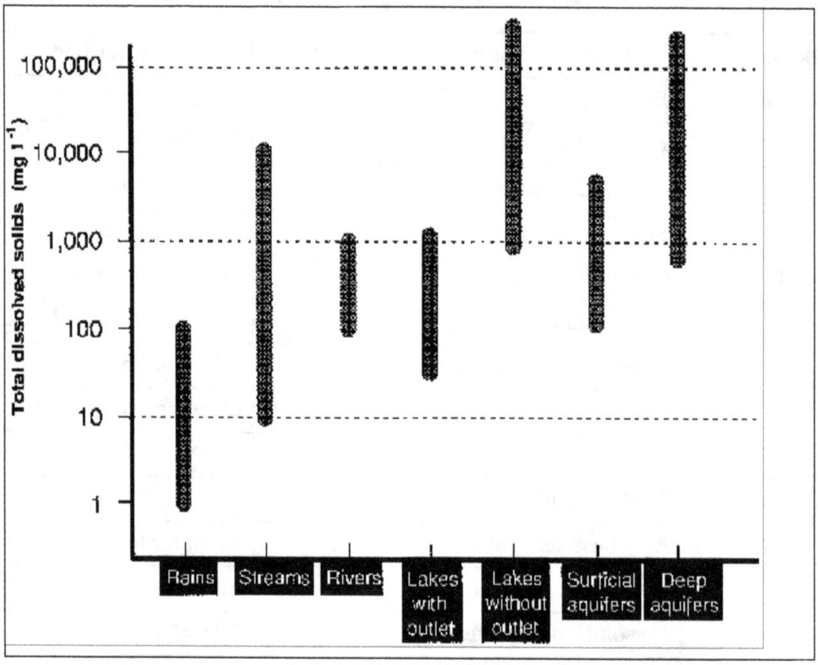

Common ranges of total dissolved solids in inland waters.

These natural variations are reflected in the range of total dissolved solids in various water bodies, as illustrated in figure. Total dissolved solids is the sum of the silica plus the major ions in the water. A close approximation of its value may be obtained by the measurement of electrical conductivity.

Some chemical elements have a strong affinity for particulate matter and, as a result of precipitation/dissolution and adsorption/desorption reactions, they may be found in only trace amounts in solution. Other elements, however, are highly soluble and rarely, if ever, present in water in particulate form. The tendency for a chemical to be present in the soluble form rather than associated with particulates is expressed as the Soluble Transport Index. Particulate matter is separated from material in solution by filtering a water sample through a filter with pore size 0.45 μm or 0.50 μm. It is recognised that some fine colloids such as the hydroxides of iron, aluminium and manganese may pass through the filter, but the use of finer filters is very costly and requires long filtration times.

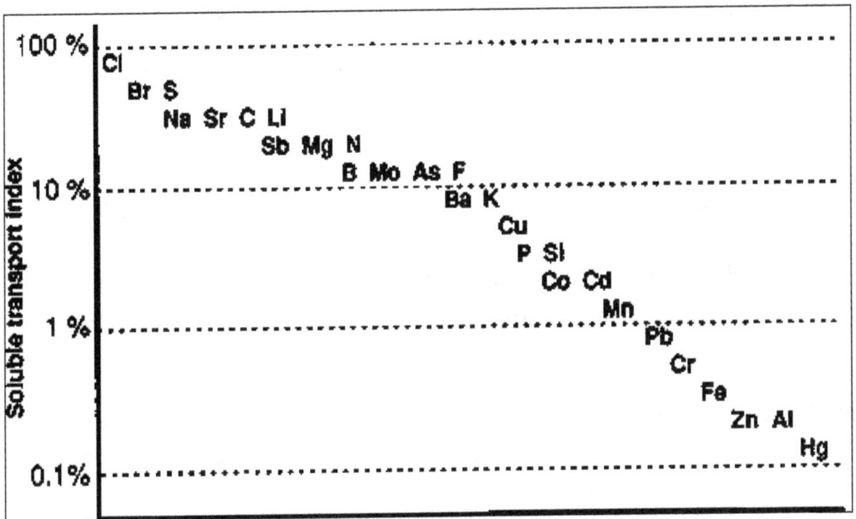

Relative scale of Soluble Transport Index (the tendency for a chemical to be present in the soluble form).

The elemental composition of river particulates is less variable than the dissolved concentrations for the same elements. This is true for the major elements aluminium, iron, silicon, calcium, magnesium, sodium, potassium, manganese, titanium and phosphorus but less so for the trace elements arsenic, cadmium, cobalt, chromium, copper, mercury, nickel, lead, selenium, tin, zinc, etc. Except for aluminium, iron, calcium and sodium, the composition of major elements in particulates is very similar in tropical rivers and in temperate and cold rivers. In small watersheds, local geological conditions can lead to wide variations in the concentration of trace elements in particulates.

Within any one water body water quality can differ with time and with place. Differences due to time are of five types:

- Minute-to-minute and day-to-day differences resulting from water mixing and fluctuations in inputs, usually as a result of meteorological conditions. These differences are most evident in small water bodies.
- Diurnal (24-hour) variations resulting from biological cycles and daylight/darkness cycles which cause changes in, for example, dissolved oxygen and pH. Diurnal patterns also result from the cyclic nature of waste discharges from domestic and industrial sources.
- Irregular patterns. Irregular sources of pollution include fertilisers, pesticides and herbicides, present in the run-off from agricultural land, and wastes discharged from food processing plants. The resultant variations in water quality may be apparent over a matter of days or months.
- Seasonal biological and hydrological cycles.
- Year-to-year trends, usually as a result of increased human activities in the watershed.

Water quality differences may result from either internal or external processes. Internal processes are usually cyclic, with either daily or seasonal recurrence, and are not directly related to the size of the water body. External processes, such as the addition of pollutants, may be buffered by large water bodies (depending on flow regimes) and long water residence times. As a result, the average composition of a very large lake probably changes little from one year to the next. Similarly, the differences in water quality at different times of the year will be much greater for a stream than for a large river. This means that the sampling frequency necessary to allow average water quality to be described correctly is normally much greater for a stream than for a river; for lakes it is normally much lower than for rivers.

Water quality differences from place to place depend more on the homogeneity of the water body than on its size. The water in a round lake, for example, may be adequately described by one sample taken from near the centre of the lake. Long, thin lakes and lakes with many bays and inlets will require more samples; the minimum is 3 while the optimum number could be 10 or more. In lakes deeper than about 30 m, or that are stratified, it is especially important to obtain samples from different depths, so that a vertical profile can be obtained. Depending on the depth of the lake, up to five samples should be taken at depths determined by conditions measured in the field.

The flow pattern, or regime, of any particular river will be the product of very specific conditions. Although similar to that of rivers in the surrounding geographical region, it will be extensively influenced by altitude, exposure of the slope to wind and variations in rainfall.

WATER QUALITY INDEX

Water Quality Index (WQI) indicates the quality of water in terms of index number which represents overall quality of water for any intended use. It is defined as a rating reflecting the composite influence of different water quality parameters were taken into consideration for the calculation of water Quality index (WQI) .The indices are among the most effective ways to communicate the information on water quality trends to the general public or to the policy makers and in water quality management. In formulation of water quality index the relative importance of various parameters depends on intended use of water. Mostly it is done from the point of view of its suitability for human consumption.

The calculation of WQI was made using weighed Arithmetic index method in the following steps:

Let there be n water quality parameters and quality rating (q_n) corresponding to nth parameter is a number reflecting relative value of this parameter in the polluted water with respect to its standard permissible value. qn values are given by the relationship.

$$q_n = 100 (v_n - v_i) / (v_s - v_i)$$

v_s = Standard value, v_n = observed value v_i = ideal value.

In most cases v_i=0 except in certain parameters like pH, dissolved oxygen etc.

Calculation of quality rating for pH & DO (vi ≠ 0):

$q_{pH} = 100 (v_{pH} - 7.0) / (8.5 - 1.0)$ and

$q_{DO} = 100 (V_{DO} - 14.6) / (15.0\text{-}14.6)$.

Calculation of unit weight : The Unit weight (Wn) to various water Quality parameters are inversely proportional to the recommended standards for the corresponding parameters.

$Wn = k/s_n$.

Where w_n = unit weight for nth parameter:

s_n = standard permissible value for nth parameter

k = proportionality constant.

WQI is calculated by the following equation.

$$WQI = \sum_{n=1}^{n} q_n w_n / \sum_{n=1}^{n} w_n$$

The suitability of WQI values for human consumption are rated as follows.

0-25 .... .Excellent

26-50 .... Good

51-75 ....Bad

76-100 .... Very Bad

100 and above .... Unfit.

Assessment of Water Quality based on WQI

Application of WQI is a useful method in assessing the suitability of water for various beneficial uses.

Temperature: Temperature of water is basically important because it effects bio-chemical reactions in aquatic organisms. A rise in temperature of water leads to the speeding up of chemical reactions in water, reduces the solubility of gases and amplifies the tastes and odours. The average temperature of the present study ranged from 26.93-28.21° c. The pH value of natural water changes due to the biological activity and industrial contamination. Higher pH includes the formation of trihalomethanes which are toxic. The pH values of the present investigation were within the ICMR standards (7.0 – 8.5).

Electrical Conductivity: Conductivity is a measure of current carrying capacity. Thus, as concentration of dissolved salts increases conductivity also increases. Many dissolved substances may produce aesthetically displeasing colour, taste and odour. The values obtained are in the range 0.9 to 1.8 mmhos.

Total Dissolve Solids: TDS values ranged within 420 to 709mg/lt. The TDS values for ground water range from 19 to 1280mg/lt as per standards and in this respect this water is suitable for drinking purposes.

Dissolved Oxygen(DO): It is an important parameter which is essential to the metabolism of all aquatic organisms that posses aerobic respiration presence of DO in water may be due to direct diffusion from air and photosynthetic activity of autotrophs. Oxygen can be rapidly removed from the waters by discharge of oxygen demanding wastes. The DO values obtained in the present study are within ICMR standards.

Alkalinity: Alkalinity value with less than 100mg/lt is desirable for domestic use. However, in large quantities imparts bitter taste to water. In the present investigation the total alkalinity of the water samples is found in the Range 106.4 to 162.2 mg/lt.

Hardness : Hardness is a measure of the ability of water to cause precipitation of insoluble calcium and magnesium salts of higher fatty acids from soap solutions. The principal hardness causing cations are calcium, magnesium bicarbonate, carbonate, chloride and sulphates. The hardness values of the present study were found to range between 220.4 to 532 mg/lit.

The quantities of calcium in natural water depend up on the type of rocks. Small concentration of calcium is beneficial in reducing the corrosion in water pipes. Magnesium hardness particularly associated with sulphate ion has laxative effect an persons un accustomed to it. In the present study calcium and magnesium contents are found in the range of 85.4 – 169.5 and 26.8- 78.2 mg/lt respectively.

Chloride: Chloride occurs in all types of natural waters. The high concentration of chloride is considered to be an indication of pollution due to high organic waste of animal origin. Chloride values obtained in the study are found in the range between 74.2-134.0 mg / lt. Bio-Chemical Oxygen Demand (BOD) & Chemical Oxygen Demand (COD) BOD & COD are the parameters used to assess the pollution of surface water and ground waters. Both of the parameters (BOD & COD) values obtained in the present study are within permissible levels.

Sulphate: Sulphate ion does not affect the taste of water, if present in low concentrations. The sulphate ion concentration in the present investigation varied from 87.5-185.0 mg/lt.

Nitrate: Nitrate is the most important nutrients in an ecosystem. Generally water bodies polluted by organic matter exhibit higher values of nitrate. In the present study water samples from the stations (s_1 to s_8) showed low concentrations of nitrate (1.34 to 2.15 mg/lt) well below permissible levels as per the standards.

WATER QUALITY MONITORING

Water quality monitoring aims to obtain qualitative information on the physical, chemical, and biological characteristics of a water body.

The objectives of a water quality monitoring program can be manifold. They range from plain monitoring of the environmental state of a water body to the detection of drinking water standard violations, from the analysis of temporary water quality trends to the observation of an industrial plant's injection of waste water. Last not least maintaining and monitoring a certain water quality level is made compulsory by many national and international regulations.

Deteriorating quality of water, for whichever reason, can pose a major hazard to human health, aquatic live and entire eco-systems. Timely detection is of crucial importance to minimize and mitigate the impact of such events, and/or to determine the source of pollution. Particularly in rivers and streams short polluting events, such as illegal disposal of toxic wastes or excessive waste water loads, are nearly impossible to detect by taking samples. Timely detection of such events can help taking preventive action as the wave of pollutants moves downstream.

The situation is aggravated by the various parameters that need to be monitored. While for many cases simple indicators such as temperature and conductivity will suffice, in other cases more sophisticated parameters, such as pH, dissolved oxygen (DO), turbidity or blue algae content will need to be monitored.

Water quality monitoring data is incredibly useful however it's not always easy to gather. Specialists use a range of different techniques to put together results, including taking samples of chemical conditions, analysing sediments and using fish tissue extracts to find traces of metals, oils, pesticides, dissolved oxygen and nutrients. Physical conditions such as temperature, erosion and flow offer valuable insight while biological measurements regarding plant and animal life indicate the health of aquatic ecosystems.

At the end of the day, water quality monitoring is an essential part of keeping the planet healthy and sustainable. As we continue to build cities, clear land for farming and make other man-made changes to the natural environment, water quality monitoring becomes increasingly important. Land based activities can have a huge impact on water systems and it's critical that we realise how these affect waterbodies, both above and below ground.

Water Quality Indicators

Water Temperature

Water temperature has a large influence on the organisms living in the water, as it influences biological activity and chemical processes. For example, water temperature influences the rate of plant photosynthesis, the timing of reproduction and migration, and the metabolic rates of aquatic organisms. Also, as the temperature of water increases, the capacity of water to hold dissolved oxygen (DO) becomes lower. Colder water will have a higher DO level than warmer water.

Runoff from paved roads, parking lots, rooftops, and other similar impervious surfaces can increase water temperature, and reduce the diversity of aquatic life.

Measuring temperature: Thermometers or probes can be used to measure water temperature.

pH

pH is a measure of how acidic or basic (alkaline) a solution is. The pH of water is critical to the survival of most aquatic plants and animals. For example, if water becomes too acidic, the eggs of some species of frogs will not develop. Most organisms are highly susceptible to changes in the pH of their surroundings or water supply, so fluctuations in pH or long-term acidification of a water body are exceedingly harmful. The pH of water can affect the pH of an organism's body fluids, can affect the speed of chemical reactions within the body, and can impact biological activities including photosynthesis, respiration, and reproduction. For example, the colorful brook trout, found in many of Virginia's streams, will experience inhibited reproduction if the pH falls below 5, and adult brook trout will die when the pH is below 4.5.

Some of the factors that determine the pH of water are bacterial activity, photosynthesis and respiration, water turbulence, humanmade wastes, acid precipitation, chemicals found in runoff, and accidental spills. Normal pH in most of Virginia's surface waters is 6.0 to 8.0, although some swamps in Virginia are naturally lower. In Virginia's mountains, many streams have low natural buffering capacity, and have been affected by acid rain. Acid precipitation in Virginia has been measured at 4.5 and lower National Atmospheric Deposition Program. This acidic precipitation has negative impacts on aquatic ecosystems, mostly in Virginia's mountainous areas due to low acid neutralization capacity associated with the bedrock type, thin soil cover, and topography.

Measuring pH: Several easy-to-use probes and test kits are available for testing the pH of water. Test paper strips to obtain pH are not suitable for use in salt water, as they will not provide consistent measurements. Schools in the coastal plain should use probes or test kits.

Transparency

Even unpolluted water can have suspended particles, such as soil particles, tiny floating organisms, or fragments of dead plants. Natural runoff, wave action, and turbulent water can cause short periods of turbid, or unclear, water. Short-term turbidity, however, doesn't have the negative impacts that are associated with long-term turbidity. Human activities, including runoff from construction sites, urban areas, and agricultural fields, can contribute to high turbidity that can lasts for long periods. Heavy boat traffic and recreational personal watercraft can also stir up sediment and increase erosion from riverbanks. High levels of turbidity over long periods of time can greatly diminish the health and productivity of aquatic ecosystems. Suspended particles can clog gills of fish and other aquatic animals. Many animals living in water feed by filtering the water, and their filter-feeding system can become impaired when suspended material is in the water in large quantities. Turbid water also reduces photosynthesis in submerged aquatic plants, as light cannot penetrate as far into the water. When the suspended particles settle out, they can cover the feeding and spawning grounds of aquatic animals, thereby destroying critical habitat.

Measuring the transparency of water: Secchi disks (a weighted black and white patterned disks on a measured rope) are widely used to measure the transparency of deep water. One should lower a Secchi diskinto the water on the shady side of a boat or dock, and record the depth at which it just disappears from view. For shallow water, as found in a stream, use a transparency tube to measure the water's transparency. Transparency tubes are sometimes referred to as turbidity tubes). Both of these are easy to use, inexpensive to purchase or make, and consistently accurate. Turbidity can also be measured with a turbidity field kit, or turbidity meter.

Nutrients

Nutrients, like nitrogen and phosphorus, that occur naturally in water, soil, and air, are essential to aquatic plants and animals, just as they are to agricultural crops, garden plants, and terrestrial animals. For example, nitrogen is used to synthesize protein in plants and animals, and phosphorus is critical in the process of photosynthesis. Both are essential to cellular growth and reproduction. In addition to natural sources, human activities contribute nutrients to the Chesapeake Bay and other water bodies. Human sources include point and nonpoint source discharges including effluent from wastewater treatment plants, urban stormwater runoff, lawn and agricultural fertilizer runoff, industrial discharges, livestock wastes, and failing septic systems. Another important source is atmospheric deposition from burning fossil fuels.

Excessive concentrations of nutrients can cause great harm to the water quality of our rivers, lakes, and the Chesapeake Bay.

Depending on their chemical forms, nitrogen and phosphorus can have significant direct or indirect impacts on plant growth, water clarity, sedimentation rates, and dissolved oxygen concentrations. As Virginia's population has grown, and land use in Virginia has changed from forests and wetlands to farms, urban areas, and suburbs, we have seen a tremendous increase in the amount of nutrients in our surface waters, including the Chesapeake Bay. In fact, the main cause of the Bay's poor water quality and aquatic habitat loss is elevated levels of nutrients.

According to the Chesapeake Bay Program, "Excess amounts of phosphorus and nitrogen cause rapid growth of phytoplankton, creating dense populations, or blooms. These blooms become so dense that they reduce the amount of sunlight available to submerged aquatic vegetation (SAV). Without sufficient light, plants cannot photosynthesize and produce the food they need to survive. The loss of sunlight can kill the grasses. Algae may also grow directly on the surface of SAV. Unconsumed algae will ultimately sink and be decomposed by bacteria in a process that depletes bottom waters of oxygen. Like humans, most aquatic species require oxygen. When oxygen in deep water is depleted, fish and other species will die unless they move to other areas of suitable habitat."

Measuring nutrients in water: Field test kits can be used to measure the concentrations of nitrates and phosphorus in a water body. Most kits rely on a color comparison in which the student matches the color of a prepared water sample to a set of provided standards.

Unpolluted water normally has less than 1 part per million (ppm) of nitrate and less than 0.03 ppm of phosphorus. The drinking water standard for nitrate nitrogen is 10 ppm. Nitrates in large amounts can cause "blue baby syndrome" (methemoglobinemia) in infants.

Dissolved Oxygen

Dissolved oxygen (DO) is important to the health of aquatic ecosystems, as all aquatic animals need oxygen for respiration. It is one of the best indicators of the water's health. Water bodies with high levels of DO are most likely healthy, and are capable of supporting a diversity of organisms. Oxygen enters water through aquatic plant photosynthesis and from the atmosphere. A riffle in a stream, where the water flows over rocks with turbulence, will boost the amount of oxygen in the water by putting more water in contact with the atmosphere. Currents and wind-generated waves will also increase the DO.

DO levels can be decreased by thermal pollution, and some pollutants like high levels of bacteria from sewage, or the rotting of plants. Oxygen is needed to aid in decomposition of organic matter. This is why an algae bloom, caused by excessive nutrients, can lead to decreased DO. Decomposition of large quantities of organic matter by bacteria can severely deplete the water of oxygen and make it uninhabitable for many species.

Removing streamside trees that shade the water will also cause the water to increase in temperature, and therefore will decrease the amount of DO the water can hold.

In addition to measuring DO, another important measurement is "DO percent saturation." Cold water can hold more DO than heated water. For example, freshwater at 8°C can hold up to 12 parts per million (ppm) of oxygen before it is 100% saturated. At 28°C, freshwater can hold only 8 ppm of DO before it is 100% saturated. Salinity of water will also affect DO levels, as freshwater can hold more DO than saltwater.

Measuring dissolved oxygen: Dissolved oxygen field test kits that measure DO to the nearest two tenths of a part per million (ppm) cost $30 to $90, and rely on the Winkler titration method. These tests usually involve more steps than the test kits for pH and nutrients, and have five chemicals that are added to the water sample in a prescribed order. Simplified test kits that measure DO at 0, 4 and 8 ppm also exist for younger children. DO meters or probes are expensive and need careful maintenance. For these reasons, most water monitoring programs use the field test kits. If you wish to measure DO percent saturation, a thermometer or temperature probe will also be needed.

Salinity

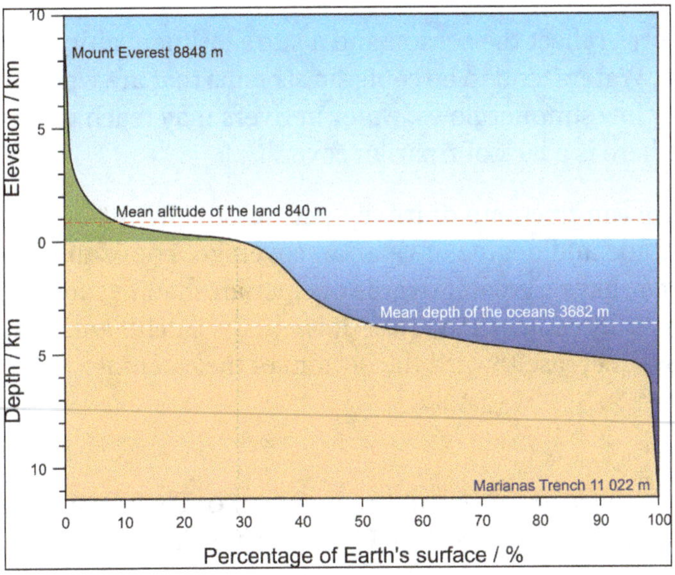

Schools near the Atlantic Ocean, the Chesapeake Bay, or tidal streams should consider adding salinity (the amount of dissolved salts in water) to their school's water monitoring program. Salinity measurements can be used to track the impacts of storm water runoff as well as the impact of drought. Salinity levels control, to a large degree, the types of plants and animals that can live in the water. Some species tolerate only fresh water (salinity of 1.0 part per thousand or less), others tolerate only brackish water (intermediate levels of salinity), while others are adapted to live only in seawater where salinity levels are 30.0 ppt.

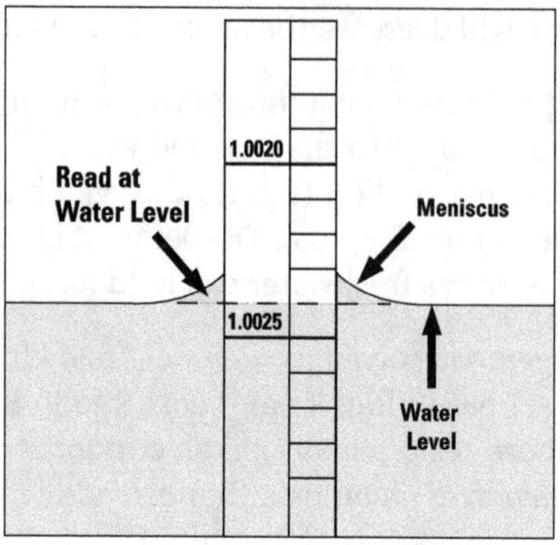

Measuring salinity: Salinity can be measured either by physical methods (conductivity, density or refractivity), or by chemical methods (measuring the chloride concentration with a field test kit). The density of water is related to the amount of salt dissolved in it, and a hydrometer measures this density, or specific gravity.

Temperature

Stream temperatures reflect the season and a site's latitude, altitude, aspect and proximity to the ocean. Waters tend to be cooler in streams that are well shaded or rise in alpine areas. During low summer flows, water in rivers may reach warmer temperatures, especially where there is a lack of riparian cover.

Water temperature can fluctuate diurnally and seasonally. Water temperature is also influenced by altitude and degree of riparian coverage. For example top of catchment locations which may have a greater degree of riparian shading, and are typically higher in altitude are generally cooler than lower down in the catchment. Basically, on average water temperatures increase toward the bottom of the catchment.

pH

The Australian guidelines for fresh waters are 6.5 - 9.0. The pH or acidity of Tasmanian streams is typically in the range 5.5 - 7.5. For humic rich Tasmanian lakes and rivers, the pH range is 4.0 to 6.5. Tasmania's inland waters are also weakly buffered i.e. they are unable to absorb large pH changes.

The majority of Tasmanian inland waters are low in alkalinity and hence poorly buffered. This is with the exception of some West Coast streams which are slightly more acidic.

In stream pH values are strongly influenced by a variety of factors relating to catchment geology, soil chemistry, vegetation and land use practices. The degree to which these factors affect pH and its buffering capacity against changing environmental conditions are inter-related with levels of acidity and alkalinity. Waters of low alkalinity (< 24 ml/L as $CaCO_3$) have a lower buffering capacity and are therefore more susceptible to fluctuations in pH. These fluctuations can be influenced by biological and atmospheric processes. Thus pH levels can vary both seasonally and diurnally depending upon environmental conditions.

The pH is a measure of the acidity of a solution and ranges in scale from 0 to 14 (from very acid to very alkaline). A pH value of 7 is considered 'neutral'. In natural waters, pH is generally between 6.0 and 8.5. In waters with little or no buffering capacity, pH is related to alkalinity which is controlled by concentrations of carbonates, bicarbonates and hydroxides in the water. Waters of low alkalinity (< 24 ml/L as $CaCO_3$) have a low buffering capacity and are susceptible to changes in pH from outside sources." (State of Rivers Report for Rivers in the Great Forester catchment).

Salinity (Electrical Conductivity)

Conductivity is a measure of the capacity of an aqueous solution to carry an electrical current, and depends on the presence of ions; on their total concentration, mobility and valence. Conductivity is commonly used to determine salinity and is mostly reported in microSiemens per centimetre (mS/cm) or milliSiemens per metre (mS/m) at a standard reference temperature of 25 degrees Celsius.

The recommended default trigger values for lowland rivers are 125-2200 mScm-1, upland rivers are 30 - 350 mScm-1, and 20 - 30 mScm-1 for lakes and reservoirs. Tasmanian streams are generally at a mid-range of 90 mScm-1.

Changes in conductivity can also be explained through landuse, catchment geology and flow. For example during periods of low flow, ground water has a greater influence on conductivity. Hence in some regions of the state (eg Jordan River catchment, Coal River catchment and some east coast catchments) where ground water is naturally saline, low summer flows are often typified by high levels of conductivity due to ground water dominance. During higher winter flows the lower saline surface waters dominate, resulting in a decrease in in-stream conductivity. In some catchments it is not unusual for an initial spike in conductivity to occur at the beginning of a high flow event. A flushing of salts from the soil profile in the immediate area usually causes this.

Turbidity

Turbidity in water is caused by suspended material such as clay, silt, finely divided organic and inorganic matter, soluable coloured compounds and plankton and microscopic organisms. Turbidity is an expression of the optical properties that cause light to be scattered and absorbed rather than transmitted in a straight line through the water. Standard units for turbidity are "nephelometric turbidity units" (NTU's) standardised against Formazin solution.

Suspended Solids

Suspended solids are typically comprised of clay, silt, fine particulate organic and inorganic matter and microscopic organisms. Suspended solids are that fraction which will not pass through a 0.45 mm filter and as such corresponds to non-filterable residues. It is this fraction which tends to contribute most to the turbidity of water.

Dissolved Oxygen

Dissolved oxygen (DO) concentrations vary with temperature, salinity, biological activity and rate of transfer from the atmosphere and provides a basic indicator of ecosystem health. Cooler waters are more capable of holding DO than warmer waters and hence there are generally higher concentrations of DO during the cooler months as opposed to the warmer summer months. Salinity also affects DO concentrations with the solubility of DO increasing with decreasing salinity. Diurnal fluctuations in DO can be quite marked with night-time concentrations lower than those experienced during daylight hours. This is in part a function of in-stream photosynthetic and respiratory activity with decreasing DO at night influenced by the respiratory activity of aquatic plants.

The decomposition of organic matter through microbial activity also consumes oxygen and can be evident at nutrient enriched sites. The impact of microbial activity on water quality can be measured by testing for biochemical oxygen demand (BOD) of water.

Heavy Metals

Heavy metal pollution is typically associated with mining activities or discharges from some industries.

Persistent toxicants (heavy metals) in water and sediments affected by heavy metal pollution can have serious affects on the aquatic ecosystem and can make water unsuitable for human consumption. Some animals can also 'bioaccumulate' metals, making them unsafe to eat.

Nutrients

Some sources of nutrients are surface water runoff from land, resuspended material from the river bed, and bank erosion. Discharges from dairy sheds and sewage treatment plants can also contribute to elevated nutrient levels if not properly managed.

The process of nutrient enrichment is called eutrophication. Excess nutrient enrichment can lead to algal blooms. Factors contributing to the establishment of a bloom can be the transportation of large quantities of nutrients and a prolonged period of calm dry weather. Once a bloom is established it can persist for long periods even during winter.

Turbidity can sometimes be correlated with nutrients, however, to get a true relationship between turbidity and phosphate there needs to be a sizeable amount of data. Increases in total phosphate and total nitrogen can sometimes be correlated with an increase in turbidity and total suspended solids measured, particularly during flood events. This is because nutrients such as phosphates often adhere to clay particles, and it is during flood events when much of the material suspended is made up of sediment, soil and detritus. This can be from surface run off from the land, from the river bed, or eroded river banks.

FRESHWATER ENVIRONMENTAL QUALITY PARAMETERS

Freshwater environmental quality parameters are the natural and man-made chemical, biological and microbiological characteristics of rivers, lakes and ground-waters, the ways they are measured and the ways that they change. The values or concentrations attributed to such parameters can be used to describe the pollution status of an environment, its biotic status or to predict the likelihood or otherwise of a particular organisms being present. Monitoring of environmental quality parameters is a key activity in managing the environment, restoring polluted environments and anticipating the effects of man-made changes on the environment.

Freshwater environmental quality parameters are those chemical, physical or biological parameters that can be used to characterise a freshwater body. Because almost all water bodies are dynamic in their composition, the relevant quality parameters are typically expressed as a range of expected concentrations.

Characterization

The first step in understanding the chemistry of freshwater is to establish the relevant concentrations of the parameters of interest. Conventionally this is done by taking representative samples of the water for subsequent analysis in a laboratory. However, in-situ monitoring using hand-held analytical equipment or using bank-side monitoring stations are also used.

Sampling

Freshwaters are surprisingly difficult to sample because they are rarely homogeneous and their quality varies during the day and during the year. In addition the most representative sampling locations are often at a distance from the shore or bank increasing the logistic complexity.

Rivers

Filling a clean bottle with river water is a very simple task, but a single sample is only representative of that point along the river the sample was taken from and at that point in time. Understanding the chemistry of a whole river, or even a significant tributary, requires prior investigation to understand how homogeneous or mixed the flow is and to determine if the quality changes during the course of a day and during the course of a year. Almost all natural rivers will have very significant patterns of change through the day and through the seasons. Water remote sensing offers a spatially continuous tool to improve understanding of spatial and temporal river water quality. Many rivers also have a very large flow that is unseen. This flows through underlying gravel and sand layers and is called hyporheic flow. How much mixing there is between the hyporheic zone and the water in the open channel will depend on a variety of factors, some of which relate to flows leaving aquifers which may have been storing water for many years.

Ground Waters

Ground waters by their very nature are often very difficult to access to take a sample. As a consequence the majority of ground-water data comes from samples taken from springs, wells, water supply bore-holes and in natural caves. In recent decades as the need to understand ground water dynamics has increased, an increasing number or monitoring boreholes have been drilled into aquifers.

Lakes

Lakes and ponds can be very large and support a complex eco-system in which environmental parameters vary widely in all three physical dimensions and with time. Large lakes in the temperate zone often stratify in the warmer months into a warmer upper layers rich in oxygen and a colder lower layer with low oxygen levels. In the autumn, falling temperatures and occasional high winds result in the mixing of the two layers into a more homogeneous whole. When stratification occurs it not only affects oxygen levels but also many related parameters such as iron, phosphate and manganese which are all changed in their chemical form by change in the redox potential of the environment.

Lakes also receive waters, often from many different sources with varying qualities. Solids from stream inputs will typically settle near the mouth of the stream and depending on a variety of factors the incoming water may float over the surface of the lake, sink beneath the surface or rapidly mix with the lake water. All of these phenomena can skew the results of any environmental monitoring unless the process are well understood.

Mixing Zones

Where two rivers meet at a confluence there exists a mixing zone. A mixing zone may be very large and extend for many miles as in the case of the Mississippi and Missouri rivers in the United States and the River Clwyd and River Elwy in North Wales. In a mixing zone water chemistry may be very variable and can be difficult to predict. The chemical interactions are not just simple mixing but may be complicated by biological processes from submerged macrophytes and by water joining the channel from the hyporheic zone or from springs draining an aquifer.

Geological Inputs

The geology that underlies a river or lake has a major impact on its chemistry. A river flowing across very ancient precambrian schists is likely to have dissolved very little from the rocks and maybe similar to de-ionised water at least in the headwaters. Conversely a river flowing through chalk hills, and especially if its source is in the chalk, will have a high concentration of carbonates and bicarbonates of Calcium and possibly Magnesium.

As a river progresses along its course it may pass through a variety of geological types and it may have inputs from aquifers that do not appear on the surface anywhere in the locality.

Atmospheric Inputs

Oxygen is probably the most important chemical constituent of surface water chemistry, as all aerobic organisms require it for survival. It enters the water mostly via diffusion at the water-air interface. Oxygen's solubility in water decreases as water temperature increases. Fast, turbulent streams expose more of the water's surface area to the air and tend to have low temperatures and thus more oxygen than slow, backwaters. Oxygen is a by-product of photosynthesis, so systems with a high abundance of aquatic algae and plants may also have high concentrations of oxygen during the day. These levels can decrease significantly during the night when primary producers switch to respiration. Oxygen can be limiting if circulation between the surface and deeper layers is poor, if the activity of animals is very high, or if there is a large amount of organic decay occurring such as following Autumn leaf-fall.

Most other atmospheric inputs come from man-made or anthropogenic sources the most significant of which are the oxides of sulphur produced by burning sulphur rich fuels such as coal and oil which give rise to acid rain. The chemistry of sulphur oxides is complex both in the atmosphere and in river systems. However the effect on the overall chemistry is simple in that it reduces the pH of the water making it more acidic. The pH change is most marked in rivers with very low concentrations of dissolved salts as these cannot buffer the effects of the acid input. Rivers downstream of major industrial conurbations are also at greatest risk. In parts of Scandinavia and West Wales and Scotland many rivers became so acidic from oxides of sulphur that most fish life was destroyed and pHs as low as pH4 were recorded during critical weather conditions.

Anthropogenic Inputs

The majority of rivers on the planet and many lakes have received or are receiving inputs from human-kind's activities. In the industrialised world, many rivers have been very seriously polluted, at least during the 19th and the first half of the 20th centuries. Although in general there has been much improvement in the developed world, there is still a great deal of river pollution apparent on the planet.

Toxicity

In most environmental situations the presence or absence of an organism is determined by a complex web of interactions only some of which will be related to measurable chemical or biological parameters. Flow rate, turbulence, inter and intra specific competition, feeding behaviour, disease, parasitism, commensalism and symbiosis are just a few of the pressures and opportunities facing any organism or population. Most chemical constituents favour some organisms and are less favourable to others. However, there are some cases where a chemical constituent exerts a toxic effect. i.e. where the concentration can kill or severely inhibit the normal functioning of the organism.

Chemical Constituents

Colour and Turbidity

Often it is the colour of freshwater or how clear or hazy the water is that is the most obvious visual characteristic. Unfortunately neither colour nor turbidity are strong indicators of the overall chemical composition of water. However both colour and turbidity reduce the amount of light penetrating the water and can have significant impact on algae and macrophytes. Some algae in particular are highly dependent on water with low colour and turbidity.

Many rivers draining high moor-lands overlain by peat have a very deep yellow brown colour caused by dissolved humic acids.

Organic Constituents

One of the principal sources of elevated concentrations of organic chemical constituents is from treated sewage.

Dissolved organic material is most commonly measured using either the Biochemical oxygen demand (BOD) test or the Chemical oxygen demand (COD) test. Organic constituents are significant in river chemistry for the effect that they have on dissolved oxygen concentration and for the impact that individual organic species may have directly on aquatic biota.

Any organic and degradable material consumes oxygen as it decomposes. Where organic concentrations are significantly elevated the effects on oxygen concentrations can be significant and as conditions get extreme the river bed may become anoxic.

Some organic constituents such as synthetic hormones, pesticides, phthalates have direct metabolic effects on aquatic biota and even on humans drinking water taken from the river. Understanding such constituents and how they can be identified and quantified is becoming of increasing importance in the understanding of freshwater chemistry.

Metals

A wide range of metals may be found in rivers from natural sources where metal ores are present in the rocks over which the river flows or in the aquifers feeding water into the river. However many rivers have an increased load of metals because of industrial activities which include mining and quarrying and the processing and use of metals.

Iron

Iron, usually as Fe^{+++} is a common constituent of river waters at very low levels. Higher iron concentrations in acidic springs or an anoxic hyporheic zone may cause visible orange/brown staining or semi-gelatinous precipitates of dense orange iron bacterial floc carpeting the river bed. Such conditions are very deleterious to most organisms and can cause serious damage in a river system.

Coal mining is also a very significant source of Iron both in mine-waters and from stocking yards of coal and from coal processing. Long abandoned mines can be a highly intractable source of high concentrations of Iron. Low levels of iron are common in spring waters emanating from deep-seated aquifers and maybe regarding as health giving springs. Such springs are commonly called Chalybeate springs and have given rise to a number of Spa towns in Europe and the United States.

Zinc

Zinc is normally associated with metal mining, especially Lead and Silver mining but is also a component pollutant associated with a variety of other metal mining activities and with Coal mining. Zinc is toxic at relatively low concentrations to many aquatic organisms. *Microregma* starts to show a toxic reaction at concentrations as low as 0.33 mg/l.

Heavy Metals

Lead and silver in river waters are commonly found together and associated with lead mining. Impacts from very old mines can be very long-lived. In the River Ystwyth in Wales for example, the effects of silver and lead mining in the 17th and 18th centuries in the headwaters still causes unacceptably high levels of Zinc and Lead in the river water right down to its confluence with the sea. Silver is very toxic even at very low concentrations but leaves no visible evidence of its contamination.

Lead is also highly toxic to freshwater organisms and to humans if the water is used as drinking water. As with Silver, Lead pollution is not visible to the naked eye. The River Rheidol in west Wales had a major series of lead mines in its headwaters until the end of the 19th century and its mine discharges and waste tips remain to this day. In 1919 - 1921 only 14 species of invertebrates were found in the lower Rheidol when Lead concentrations were between 0.2ppm and 0.5ppm. By 1932 the lead concentration had reduced to 0.02ppm to 0.1ppm because of the abandonment of mining and, at those concentrations, the bottom fauna had stabilized to 103 species including three leeches.

Coal mining is also a very significant source of metals, especially Iron, Zinc and Nickel particularly where the coal is rich if pyrites which oxidises on contact with the air producing a very acidic leachate which is able to dissolve metals from the coal.

Significant levels of copper are unusual in rivers and where it does it occur the source is most likely to be mining activities, coal stocking, or pig farming. Rarely elevated levels may be of geological origin. Copper is acutely toxic to many freshwater organisms, especially algae, at very low concentrations and significant concentration in river water may have serious adverse effects on the local ecology.

Nitrogen

Nitrogenous compounds have a variety of sources including washout of oxides of nitrogen from the atmosphere, some geological inputs and some from macrophyte and algal nitrogen fixation. However, for many rivers in the proximity of humans, the largest input is from sewage whether treated or untreated. The nitrogen derives from breakdown products of proteins found in urine and faeces. These products, being very soluble, often pass through sewage treatment process and are discharged into rivers as a component of sewage treatment effluent. Nitrogen may be in the form of nitrate, nitrite, ammonia or ammonium salts or what is termed albuminoid nitrogen or nitrogen still within an organic proteinoid molecule.

The differing forms of nitrogen are relatively stable in most river systems with nitrite slowly transforming into nitrate in well oxygenated rivers and ammonia transforming into nitrite/nitrate. However, the process are slow in cool rivers and reduction in concentration may more often be attributed to simple dilution. All forms of nitrogen are taken up by macrophytes and algae and elevated levels of nitrogen are often associated with overgrowths of plants or eutrophication. These can have the effect of blocking channels and inhibiting navigation. However, ecologically, the more significant effect is on dissolved oxygen concentrations which may become super-saturated during daylight due to plant photosynthesis but then drop to very low levels during darkness as plant respiration uses up the dissolved oxygen. Coupled with the release of oxygen in photosynthesis is the creation of bi-carbonate ions which cause a steep rise in pH and this is matched in darkness as carbon dioxide is released through respiration which substantially lowers the pH. Thus high levels of nitrogenous compounds tends to lead to eutrophication with extreme variations in parameters which in turn can substantially degrade the ecological worth of the watercourse.

Ammonium ions also have a toxic effect, especially on fish. The toxicity of ammonia is dependent on both pH and temperature and an added complexity is the buffering effect of the blood/water interface across the gill membrane which masks any additional toxicity over about pH 8.0. The management of river chemistry to avoid ecological damage is particularly difficult in the case of ammonia as a wide range of potential scenarios of concentration, pH and temperature have to be considered and the diurnal pH fluctuation caused by photosynthesis considered. On warm summer days with high-bi-carbonate concentrations unexpectedly toxic conditions can be created.

Phosphorus

Phosphorus compounds are usually found as relatively insoluble phosphates in river water and, except in some exceptional circumstances, their origin is agriculture or human sewage. Phosphorus can encourage excessive growths of plants and algae and contribute to eutrophication. If a river discharges into a lake or reservoir phosphate can be mobilised year after year by natural processes. In the summer time, lakes stratify so that warm oxygen rich water floats on top of cold oxygen poor water. In the warm upper layers - the epilimnion- plants consume the available phosphate. As the plants die in the late summer they fall into the cool water layers underneath - the hypolimnion - and decompose. During winter turn-over, when a lake becomes fully mixed through the action of winds on a cooling body of water - the phosphates are spread throughout the lake again to feed a new generation of plants. This process is one of the principal causes of persistent algal blooms at some lakes.

Arsenic

Geological deposits of arsenic may be released into rivers where deep ground-waters are exploited as in parts of Pakistan. Many metalloid ores such as lead, gold and copper contain traces of arsenic and poorly stored tailings may result in arsenic entering the hydrological cycle.

Solids

Inert solids are produced in all montane rivers as the energy of the water helps grind away rocks into gravel, sand and finer material. Much of this settles very quickly and provides an important substrate for many aquatic organisms. Many salmonid fish require beds of gravel and sand in which to lay their eggs. Many other types of solids from agriculture, mining, quarrying, urban run-off and sewage may block-out sunlight from the river and may block interstices in gravel beds making them useless for spawning and supporting insect life.

Bacterial, Viral and Parasite Inputs

Both agriculture and sewage treatment produce inputs into rivers with very high concentrations of bacteria and viruses including a wide range of pathogenic organisms. Even in areas with little human activity significant levels of bacteria and viruses can be detected originating from fish and aquatic mammals and from animals grazing near rivers such as deer. Upland waters draining areas frequented by sheep, goats or deer may also harbour a variety of opportunistic human parasites such as liver fluke. Consequently, there are very few rivers from which the water is safe to drink without some form of sterilisation or disinfection. In rivers used for contact recreation such as swimming, safe levels of bacteria and viruses can be established based on risk assessment.

Under certain conditions bacteria can colonise freshwaters occasionally making large rafts of filamentous mats known as *sewage fungus* – usually *Sphaerotilus natans*. The presence of such organisms is almost always an indicator of extreme organic pollution and would be expected to be matched with low dissolved oxygen concentrations and high BOD vales.

E. coli bacteria have been commonly found in recreational waters and their presence is used to indicate the presence of recent faecal contamination, but E. coli presence may not be indicative of human waste. *E. coli* are found in all warm-blooded animals. *E. coli* have also been found in fish and turtles. Enterobacteria may also persist in the environment in mud, sediments, sand and soil for considerable lengths of time.

pH

pH in rivers is affected by the geology of the water source, atmospheric inputs and a range of other chemical contaminants. pH is only likely to become an issue on very poorly buffered upland rivers where atmospheric sulphur and nitrogen oxides may very significantly depress the pH as low as pH4 or in eutrophic alkaline rivers where photosynthetic bi-carbonate ion production in photosynthesis may drive the pH up above pH10.

ANALYSIS OF WATER CHEMISTRY

Water chemistry analyses are carried out to identify and quantify the chemical components and properties of water samples. The type and sensitivity of the analysis depends on the purpose of the analysis and the anticipated use of the water. Chemical water analysis is carried out on water used in industrial processes, on waste-water stream, on rivers and stream, on rainfall and on the sea. In all cases the results of the analysis provides information that can be used to make decisions or to provide re-assurance that conditions are as expected. The analytical parameters selected are chosen to be appropriate for the decision making process or to establish acceptable normality. Water chemistry analysis is often the groundwork of studies of water quality, pollution, hydrology and geothermal waters. Analytical methods routinely used can detect and measure all the natural elements and their inorganic compounds and a very wide range of organic chemical species using methods such as gas chromatography and mass spectrometry. In water treatment plants producing drinking water and in some industrial processes using products with distinctive taste and odours, specialised organoleptic methods may be used to detect smells at very low concentrations.

Environmental Water

An EPA scientist samples water in Florida Everglades.

Samples of water from the natural environment are routinely taken and analysed as part of a pre-determined monitoring programme by regulatory authorities to ensure that waters remain unpolluted, or if polluted, that the levels of pollution are not increasing or are falling in line with an agreed remediation plan. An example of such a scheme is the Harmonised monitoring scheme operated on all the major river systems in the UK. The parameters analysed will be highly dependent on nature of the local environment and/or the polluting sources in the area. In many cases the parameters will reflect the national and local water quality standards determined by law or other regulations. Typical parameters for ensuring that unpolluted surface waters remain within acceptable chemical standards include pH, major cations and anions including Ammonia, Nitrate, Nitrite, Phosphate, Conductivity, COD, Phenol and BOD.

Drinking Water Supplies

Surface or ground water abstracted for the supply of drinking water must be capable of meeting rigorous chemical standards following treatment. This requires a detailed knowledge of the water entering the treatment plant. In addition to the normal suite of environmental chemical parameters, other parameters such as hardness, Phenol, Oil and in some cases a real-time organic profile of the incoming water as in the River Dee regulation scheme.

Industrial Process Water

In industrial process, the control of the quality of process water can be critical to the quality of the end product. Water is often used as a carrier of reagents and the loss of reagent to product must be continuously monitored to ensure that correct replacement rate. Parameters measured relate specifically to the process in use and to any of the expected contaminants that may arise as by-products. This may include unwanted organic chemicals appearing in an inorganic chemical process through contamination with oils and greases from machinery. Monitoring the quality of the waste water discharged from industrial premises is a key factor in controlling and minimising pollution of the environment. In this application monitoring schemes analyse for all possible contaminants arising within the process and in addition contaminants that may have particularly adverse impacts on the environment such as Cyanide and many organic species such as pesticides. In then nuclear industry analysis focuses on specific isotopes or elements of interest. Where the nuclear industry makes waste water discharges to rivers which have drinking water abstraction on them, radio-isotopes which could potentially be harmful or those with long half-lives such as Tritium will form part of the routine monitoring suite.

Forensic Analysis

In environmental management, water analysis is frequently deployed when pollution is suspected to identify the pollutant in order to take remedial action. The analysis can often enable the polluter to be identified. Such forensic work can examine the ratios of various components and can "type" samples of oils or other mixed organic contaminants to directly link the pollutant with the source. In drinking water supplies the cause of unacceptable quality can similarly be determined by carefully targeted chemical analysis of samples taken throughout the distribution system. In manufacturing, off-spec products may be directly tied back to unexpected changes in wet processing stages and analytical chemistry can identify which stages may be at fault and for what reason.

Methodology

To ensure consistency and repeatability, the methods use in the chemical analysis of water samples are often agreed and published at a national or state level. By convention these are often referred to as "Blue book"s.

The methods defined in the relevant standards can be broadly classified as:

- Conventional wet chemistry including Winkler method for dissolved oxygen, precipitation, filtration for solids, acidification, neutralisation titration etc. Colourimetric methods such as MBAS assay which indicates anionic surfactants in water and on site comparator methods to determine Chlorine and Chloramines. Nephelometers are used to measure solids concentrations as turbidity. These methods are generally robust and well tried and inexpensive, giving a reasonable degree of accuracy at modest sensitivity.

- Electro chemistry including pH, Conductivity and dissolved oxygen using oxygen electrode. These methods yield accurate and precise results using electronic equipment capable of feeding results directly into a laboratory data management system.
- Spectrophotometry particularly for metallic elements in solution producing results with very high sensitivity but which may require some sample preparation prior to analysis and may also need specialised sampling methods to avoid sample deterioration in transit.
- Chromatography is used for many organic species which are volatile or which can yield a characteristic volatile component of after initial chemical processing.
- Ion chromatography is a sensitive and stable technique that can measure Li, NH_4 and many other low molecular weight ions using ion exchange technology.
- Gas chromatography can be used to determine methane, carbon dioxide, Cyanide oxygen and nitrogen and many other volatile components at reasonable sensitivities.
- Mass spectrometry is used where very high sensitivity is required and is sometimes used as a back-end process after Gas liquid chromatography for detecting trace organic chemicals.

Depending on the components, different methods are applied to determine the quantities or ratios of the components. While some methods can be performed with standard laboratory equipment, others require advanced devices, such as Inductively coupled plasma mass spectrometry (ICP-MS).

BACTERIOLOGICAL WATER ANALYSIS

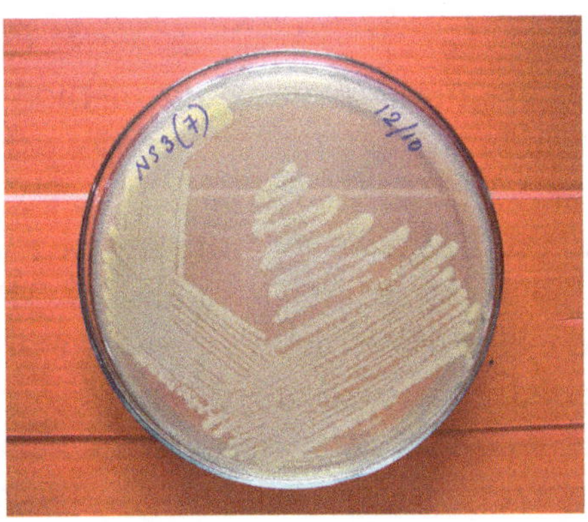

E. coli culture on a Petri dish.

Bacteriological water analysis is a method of analysing water to estimate the numbers of bacteria present and, if needed, to find out what sort of bacteria they are. It represents one aspect of water quality. It is a microbiological analytical procedure which uses samples of water and from these samples determines the concentration of bacteria. It is then possible to draw inferences about the suitability of the water for use from these concentrations. This process is used, for example, to routinely confirm that water is safe for human consumption or that bathing and recreational waters are safe to use.

The interpretation and the action trigger levels for different waters vary depending on the use made of the water. Whilst very stringent levels apply to drinking water, more relaxed levels apply to marine bathing waters, where much lower volumes of water are expected to be ingested by users.

Approach

The common feature of all these routine screening procedures is that the primary analysis is for indicator organisms rather than the pathogens that might cause concern. Indicator organisms are bacteria such as non-specific coliforms, *Escherichia coli* and *Pseudomonas aeruginosa* that are very commonly found in the human or animal gut and which, if detected, may suggest the presence of sewage. Indicator organisms are used because even when a person is infected with a more pathogenic bacteria, they will still be excreting many millions times more indicator organisms than pathogens. It is therefore reasonable to surmise that if indicator organism levels are low, then pathogen levels will be very much lower or absent. Judgements as to suitability of water for use are based on very extensive precedents and relate to the probability of any sample population of bacteria being able to be infective at a reasonable statistical level of confidence.

Analysis is usually performed using culture, biochemical and sometimes optical methods. When indicator organisms levels exceed pre-set triggers, specific analysis for pathogens may then be undertaken and these can be quickly detected (where suspected) using specific culture methods or molecular biology.

Methodologies

The most reliable methods are direct plate count method and membrane filtration method. mEndo Agar is used in the membrane filtration while VRBA Agar is used in the direct plate count method. VRBA stands for violet red bile agar. A media that contains bile salts which promotes the growth of gram negative and has inhibitory characteristic to gram positive although not complete inhibitory. These media contain lactose which is usually fermented by lactose fermenting bacteria producing colonies that can be identified and characterised. Lactose fermenting produce colored colonies while non lactose fermenting produce colorless ones. Because the analysis is always based on a very small sample taken from a very large volume of water, all methods rely on statistical principles.

Multiple Tube Method

One of the oldest methods is called the multiple tube method. In this method a measured sub-sample (perhaps 10 ml) is diluted with 100 ml of sterile growth medium and an aliquot of 10 ml is then decanted into each of ten tubes. The remaining 10 ml is then diluted again and the process repeated. At the end of 5 dilutions this produces 50 tubes covering the dilution range of 1:10 through to 1:10000.

The tubes are then incubated at a pre-set temperature for a specified time and at the end of the process the number of tubes with growth in is counted for each dilution. Statistical tables are then used to derive the concentration of organisms in the original sample. This method can be enhanced by using indicator medium which changes colour when acid forming species are present and by including a tiny inverted tube called a Durham tube in each sample tube. The Durham inverted tube catches any gas produced. The production of gas at 37 degrees Celsius is a strong indication of the presence of *Escherichia coli*.

ATP Testing

An ATP test is the process of rapidly measuring active microorganisms in water through detection adenosine triphosphate (ATP). ATP is a molecule found only in and around living cells, and as such it gives a direct measure of biological concentration and health. ATP is quantified by measuring the light produced through its reaction with the naturally occurring enzyme firefly luciferase using a luminometer. The amount of light produced is directly proportional to the amount of biological energy present in the sample.

Second generation ATP tests are specifically designed for water, wastewater and industrial applications where, for the most part, samples contain a variety of components that can interfere with the ATP assay.

Plate Count

The plate count method relies on bacteria growing a colony on a nutrient medium so that the colony becomes visible to the naked eye and the number of colonies on a plate can be counted. To be effective, the dilution of the original sample must be arranged so that on average between 30 and 300 colonies of the target bacterium are grown. Fewer than 30 colonies makes the interpretation statistically unsound whilst greater than 300 colonies often results in overlapping colonies and imprecision in the count. To ensure that an appropriate number of colonies will be generated several dilutions are normally cultured. This approach is widely utilised for the evaluation of the effectiveness of water treatment by the inactivation of representative microbial contaminants such as *E. coli* following ASTM D5465.

The laboratory procedure involves making serial dilutions of the sample (1:10, 1:100, 1:1000, etc.) in sterile water and cultivating these on nutrient agar in a dish that is sealed and incubated. Typical media include plate count agar for a general count or MacConkey agar to count Gram-negative bacteria such as *E. coli*. Typically one set of plates is incubated at 22 °C and for 24 hours and a second set at 37 °C for 24 hours. The composition of the nutrient usually includes reagents that resist the growth of non-target organisms and make the target organism easily identified, often by a colour change in the medium. Some recent methods include a fluorescent agent so that counting of the colonies can be automated. At the end of the incubation period the colonies are counted by eye, a procedure that takes a few moments and does not require a microscope as the colonies are typically a few millimetres across.

Membrane Filtration

Most modern laboratories use a refinement of total plate count in which serial dilutions of the sample are vacuum filtered through purpose made membrane filters and these filters are themselves laid on nutrient medium within sealed plates. The methodology is otherwise similar to conventional total plate counts. Membranes have a printed millimetre grid printed on and can be reliably used to count the number of colonies under a binocular microscope.

Pour Plate Method

When the analysis is looking for bacterial species that grow poorly in air, the initial analysis is done by mixing serial dilutions of the sample in liquid nutrient agar which is then poured into bottles which are then sealed and laid on their sides to produce a sloping agar surface. Colonies that develop in the body of the medium can be counted by eye after incubation.

The total number of colonies is referred to as the total viable count (TVC). The unit of measurement is cfu/ml (or colony forming units per millilitre) and relates to the original sample. Calculation of this is a multiple of the counted number of colonies multiplied by the dilution used.

Pathogen Analysis

When samples show elevated levels of indicator bacteria, further analysis is often undertaken to look for specific pathogenic bacteria. Species commonly investigated in the temperate zone include *Salmonella typhi* and *Salmonella Typhimurium*. Depending on the likely source of contamination investigation may also extend to organisms such as *Cryptosporidium spp*. In tropical areas analysis of *Vibrio cholerae* is also routinely undertaken.

Types of Nutrient Media used in Analysis

MacConkey agar is culture medium designed to grow Gram-negative bacteria and stain them for lactose fermentation. It contains bile salts (to inhibit most Gram-positive bacteria), crystal violet dye (which also inhibits certain Gram-positive bacteria), neutral red dye (which stains microbes fermenting lactose), lactose and peptone. Alfred Theodore MacConkey developed it while working as a bacteriologist for the Royal Commission on Sewage Disposal in the United Kingdom.

Endo agar contains peptone, lactose, dipotassium phosphate, agar, sodium sulfite, basic fuchsin and was originally developed for the isolation of *Salmonella typhi*, but is now commonly used in water analysis. As in MacConkey agar, coliform organisms ferment the lactose, and the colonies become red. Non-lactose-fermenting organisms produce clear, colourless colonies against the faint pink background of the medium.

mFC medium is used in membrane filtration and contains selective and differential agents. These include rosolic acid to inhibit bacterial growth in general, except for faecal coliforms, bile salts inhibit non-enteric bacteria and aniline blue indicates the ability of faecal coliforms to ferment lactose to acid that causes a pH change in the medium.

TYEA medium contains tryptone, yeast extract, common salt and L-arabinose per liter of glass distilled water and is a non selective medium usually cultivated at two temperatures (22 and 36 °C) to determine a general level of contamination (a.k.a. colony count).

AQUATIC BIOMONITORING

Aquatic biomonitoring is the science of inferring the ecological condition of rivers, lakes, streams, and wetlands by examining the organisms that live there. While aquatic biomonitoring is the most common form of biomonitoring, any ecosystem can be studied in this manner.

Biomonitoring typically takes different approaches:

- Bioassays, where test organisms are exposed to an environment to see if mutations or deaths occur. Typical organisms used in bioassays are fish, water fleas (Daphnia), and frogs.
- Community assessments, also called *biosurveys,* where an entire community of organisms is sampled to see what types of taxa remain. In aquatic ecosystems, these assessments often focus on invertebrates, algae, macrophytes (aquatic plants), fish, or amphibians. Rarely, other large vertebrates (reptiles, birds, and mammals) may be considered as well.
- Online biomonitoring devices, using the ability of animals to permanently taste their environment. Different types of animals are used for this purpose either in the lab or in the field. The study of the opening and closing activity of clams'

valves is an example of one possible way to monitor *in-situ* the quality of fresh and coastal waters.

A biosurvey on the North Toe River. North Carolina.

Aquatic invertebrates have the longest history of use in biomonitoring programs. In typical unpolluted temperate streams of Europe and North America, certain insect taxa predominate. Mayflies (Ephemeroptera), caddisflies (Trichoptera), and stoneflies (Plecoptera) are the most common insects in these undisturbed streams. In contrast, in rivers disturbed by urbanization, agriculture, forestry, and other perturbations, flies (Diptera), and especially midges (family Chironomidae) predominate. Aquatic invertebrates are responsive to climate change.

Importance

Aquatic biomonitoring is important in assessing marine life forms and their ecosystems. Monitoring aquatic life, from which life on land evolved, can also be beneficial in understanding land ecosystems.

Aquatic biomonitoring can reveal the overall health and status of the environment, can detect environmental trends and how different stressors will affect those trends, and can interpret the effect that various environmental activities will have on the overall health of the environment. Pollution and general stresses to aquatic life can have a major impact on the environment. The main sources of pollution to oceans, rivers, and lakes are sewage, oil spills, land runoff, littering, ocean mining, and nuclear waste. Pollution greatly upsets marine life and can endanger species that live in or close to water. Because many aquatic animals serve as a main food source for many land animals, when aquatic species are affected, it causes a ripple effect in land species. Biomonitoring can help mitigate such problems through monitoring all forms of life and conditions in different bodies of water, both fresh and salt water.

A challenge in aquatic biomonitoring is to simplify data and make data easier for all to understand, especially investigators in the health and environmental fields.

Practice

Methods employed in aquatic biomonitoring are:

- Monitoring and assessing aquatic species and ecosystems.
- Monitoring the behavior of certain aquatic species and assessing any changes in species behavior.
- Looking at contaminants in the water and their effect on marine life.

Water quality is graded both on appearance—for example: clear, cloudy, full of algae—and on its chemistry levels. Determining levels of enzymes and minerals found in water is extremely important. Changes in these factors can change the overall aquatic environment and can severely impact aquatic life. Some contaminants, such as metal and certain organic waste, can be lethal to individual creatures and could thereby ultimately lead to extinction of certain species. This could affect both aquatic and land ecosystems and cause disruption in other biomes and ecosystems.

WATER POLLUTION

Raw sewage and industrial waste in the New River as it passes from Mexicali (Mexico) to Calexico, California.

Water pollution is the contamination of water bodies, usually as a result of human activities. Water bodies include for example lakes, rivers, oceans, aquifers and groundwater. Water pollution results when contaminants are introduced into the natural environment. For example, releasing inadequately treated wastewater into natural water bodies can lead to degradation of aquatic ecosystems. In turn, this can lead to public health problems for people living downstream. They may use the same polluted river water for drinking or bathing or irrigation. Water pollution is the leading worldwide cause of death and disease, e.g. due to water-borne diseases.

Water pollution can be grouped into surface water pollution. Marine pollution and nutrient pollution are subsets of water pollution. Sources of water pollution are either point sources and non-point sources. Point sources have one identifiable cause of the pollution, such as a storm drain, wastewater treatment plant or stream. Non-point sources are more diffuse, such as agricultural runoff. Pollution is the result of the cumulative effect over time. All plants and organisms living in or being exposed to polluted water bodies can be impacted. The effects can damage individual species and impact the natural biological communities they are part of.

The causes of water pollution include a wide range of chemicals and pathogens as well as physical parameters. Contaminants may include organic and inorganic substances. Elevated temperatures can also lead to polluted water. A common cause of thermal pollution is the use of water as a coolant by power plants and industrial manufacturers. Elevated water temperatures decrease oxygen levels, which can kill fish and alter food chain composition, reduce species biodiversity, and foster invasion by new thermophilic species.

Water pollution is measured by analysing water samples. Physical, chemical and biological tests can be done. Control of water pollution requires appropriate infrastructure and management plans. The infrastructure may include wastewater treatment plants. Sewage treatment plants and industrial wastewater treatment plants are usually required to protect water bodies from untreated wastewater. Agricultural wastewater treatment for farms, and erosion control from construction sites can also help prevent water pollution. Nature-based solutions are another approach to prevent water pollution. Effective control of urban runoff includes reducing speed and quantity of flow. In the United States, best management practices for water pollution include approaches to reduce the quantity of water and improve water quality.

Pollution in the Lachine Canal, Canada.

Water is typically referred to as polluted when it is impaired by anthropogenic contaminants. Due to these contaminants it either does not support a human use, such as drinking water, or undergoes a marked shift in its ability to support its biotic communities, such as fish. Natural phenomena such as volcanoes, algae blooms, storms, and earthquakes also cause major changes in water quality and the ecological status of water.

Water pollution is a major global problem. It requires ongoing evaluation and revision of water resource policy at all levels (international down to individual aquifers and wells). It has been suggested that water pollution is the leading worldwide cause of death and diseases. Water pollution accounted for the deaths of 1.8 million people in 2015.

The organization Global Oceanic Environmental Survey (GOES) consider water pollution as one of the main environmental problems that can present a danger for the existence of life on earth in the next decades. One of the main concerns, is that water pollution, heart phytoplankton who produce 70% of oxygen and remove a large part of carbon dioxide on earth. The organization proposes a number of measures for fixing the situation, but they should be taken in the next 10 years for being effective.

India and China are two countries with high levels of water pollution: An estimated 580 people in India die of water pollution related illness (including waterborne diseases) every day. About 90 percent of the water in the cities of China is polluted. As of 2007, half a billion Chinese had no access to safe drinking water.

In addition to the acute problems of water pollution in developing countries, developed countries also continue to struggle with pollution problems. For example, in a report on water quality in the United States in 2009, 44 percent of assessed stream miles, 64 percent of assessed lake acres, and 30 percent of assessed bays and estuarine square miles were classified as polluted.

Types

Surface Water Pollution

Surface water pollution includes pollution of rivers, lakes and oceans. A subset of surface water pollution is marine pollution.

Mean eutrophying emissions (water pollution) of different foods per 100g of protein	
Food Types	Eutrophying Emissions (g PO_4^{3-}eq per 100g protein)
Beef	365.3
Farmed Fish	235.1
Farmed Crustaceans	227.2
Cheese	98.4

Lamb and Mutton	97.1
Pork	76.4
Poultry	48.7
Eggs	21.8
Groundnuts	14.1
Peas	7.5
Tofu	6.2

Marine Pollution

A polluted river draining an abandoned copper mine on Anglesey.

One common path of entry by contaminants to the sea are rivers. An example is directly discharging sewage and industrial waste into the ocean. Pollution such as this occurs particularly in developing nations. In fact, the 10 largest emitters of oceanic plastic pollution worldwide are, from the most to the least, China, Indonesia, Philippines, Vietnam, Sri Lanka, Thailand, Egypt, Malaysia, Nigeria, and Bangladesh, largely through the rivers Yangtze, Indus, Yellow, Hai, Nile, Ganges, Pearl, Amur, Niger, and the Mekong, and accounting for "90 percent of all the plastic that reaches the world's oceans."

Large gyres (vortexes) in the oceans trap floating plastic debris. Plastic debris can absorb toxic chemicals from ocean pollution, potentially poisoning any creature that eats it. Many of these long-lasting pieces end up in the stomachs of marine birds and animals. This results in obstruction of digestive pathways, which leads to reduced appetite or even starvation.

There are a variety of secondary effects stemming not from the original pollutant, but a derivative condition. An example is silt-bearing surface runoff, which can inhibit the penetration of sunlight through the water column, hampering photosynthesis in aquatic plants.

Groundwater Pollution

Interactions between groundwater and surface water are complex. Consequently, groundwater pollution, also referred to as groundwater contamination, is not as easily classified as surface water pollution. By its very nature, groundwater aquifers are susceptible to contamination from sources that may not directly affect surface water bodies. The distinction of point vs. non-point source may be irrelevant.

Analysis of groundwater contamination may focus on soil characteristics and site geology, hydrogeology, hydrology, and the nature of the contaminants. Causes of groundwater pollution include: naturally-occurring (geogenic), on-site sanitation systems, sewage, fertilizers and pesticide, commercial and industrial leaks, hydraulic fracturing, landfill leachate.

Categories of Pollution Sources

Surface water and groundwater have often been studied and managed as separate resources even though they are interrelated. Surface water seeps through the soil and becomes groundwater. Conversely, groundwater can also feed surface water sources. Sources of surface water pollution are generally grouped into two categories based on their origin.

Point Sources

Point source pollution at a shipyard in Rio de Janeiro, Brazil.

Point source water pollution refers to contaminants that enter a waterway from a single, identifiable source, such as a pipe or ditch. Examples of sources in this category include discharges from a sewage treatment plant, a factory, or a city storm drain.

The U.S. Clean Water Act (CWA) defines point source for regulatory enforcement purposes. The CWA definition of point source was amended in 1987 to include municipal storm sewer systems, as well as industrial storm water, such as from construction sites.

Non-point Sources

Blue drain and yellow fish symbol used by the UK Environment Agency to raise awareness of the ecological impacts of contaminating surface drainage.

Nonpoint source pollution refers to diffuse contamination that does not originate from a single discrete source. This type of pollution is often the cumulative effect of small amounts of contaminants gathered from a large area. A common example is the leaching out of nitrogen compounds from fertilized agricultural lands. Nutrient runoff in storm water from "sheet flow" over an agricultural field or a forest are also cited as examples of non-point source pollution.

Contaminated storm water washed off of parking lots, roads and highways, called urban runoff, is sometimes included under the category of non-point sources. This runoff becomes a point source because it is typically channeled into storm drain systems and discharged through pipes to local surface waters.

Contaminants and their Sources

The specific contaminants leading to pollution in water include a wide spectrum of chemicals, pathogens, and physical changes such as elevated temperature and discoloration. While many of the chemicals and substances that are regulated may be naturally occurring (calcium, sodium, iron, manganese, etc.) the concentration usually determines what is a natural component of water and what is a contaminant. High concentrations of naturally occurring substances can have negative impacts on aquatic flora and fauna.

Oxygen-depleting substances may be natural materials such as plant matter (e.g. leaves and grass) as well as man-made chemicals. Other natural and anthropogenic substances may cause turbidity (cloudiness) which blocks light and disrupts plant growth, and clogs the gills of some fish species.

Alteration of water's physical chemistry includes acidity (change in pH), electrical conductivity, temperature, and eutrophication. Eutrophication is an increase in the concentration of chemical nutrients in an ecosystem to an extent that increases the primary productivity of the ecosystem. Depending on the degree of eutrophication, subsequent negative environmental effects such as anoxia (oxygen depletion) and severe reductions in water quality may occur, affecting fish and other animal populations.

Pathogens

Fecal sludge collected from pit latrines is dumped into a river at the Korogocho slum in Nairobi, Kenya.

A manhole cover unable to contain a sanitary sewer overflow.

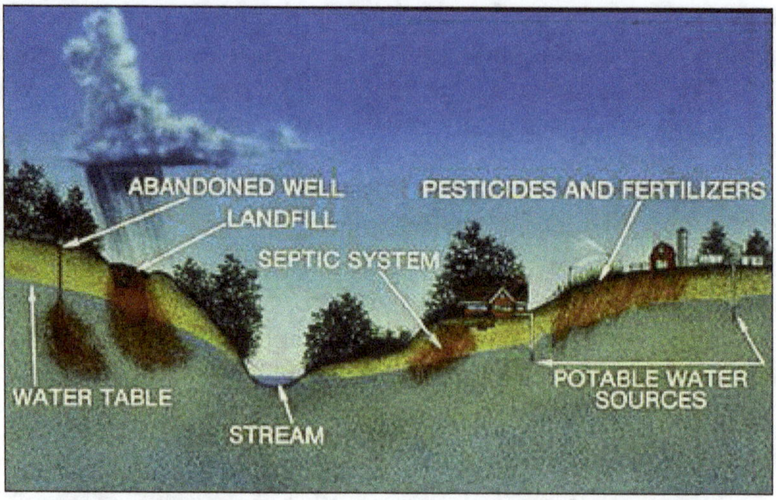

Poster to teach people in South Asia about human activities leading to the pollution of water sources.

Disease-causing microorganisms are referred to as pathogens. Pathogens can produce waterborne diseases in either human or animal hosts. Coliform bacteria, which are not an actual cause of disease, are commonly used as a bacterial indicator of water pollution. Other microorganisms sometimes found in contaminated surface waters that have caused human health problems include:

- Burkholderia pseudomallei.
- Cryptosporidium parvum.
- Giardia lamblia.
- Salmonella.
- Norovirus and other viruses.
- Parasitic worms including the Schistosoma type.

Muddy river polluted by sediment.

High levels of pathogens may result from on-site sanitation systems (septic tanks, pit latrines) or inadequately treated sewage discharges. Older cities with ageing infrastructure may have leaky sewage collection systems (pipes, pumps, valves), which can cause sanitary sewer overflows. Some cities also have combined sewers, which may discharge untreated sewage during rain storms. Silt (sediment) from sewage discharges also pollutes water bodies.

Pathogen discharges may also be caused by poorly managed livestock operations.

Organic, Inorganic and Macroscopic Contaminants

Contaminants may include organic and inorganic substances. Many of the chemical substances are toxic.

A garbage collection boom in an urban-area stream in Auckland, New Zealand.

Organic water pollutants include:

- Detergents.
- Disinfection by-products found in chemically disinfected drinking water, such as chloroform.
- Food processing waste, which can include oxygen-demanding substances, fats and grease.
- Insecticides and herbicides, a huge range of organohalides and other chemical compounds.
- Petroleum hydrocarbons, including fuels (gasoline, diesel fuel, jet fuels, and fuel oil) and lubricants (motor oil), and fuel combustion byproducts, from storm water runoff.
- Volatile organic compounds, such as industrial solvents, from improper storage.

- Chlorinated solvents, which are dense non-aqueous phase liquids, may fall to the bottom of reservoirs, since they don't mix well with water and are denser:
    - Polychlorinated biphenyl (PCBs).
    - Trichloroethylene.
- Perchlorate.
- Various chemical compounds found in personal hygiene and cosmetic products.
- Drug pollution involving pharmaceutical drugs and their metabolites, this can include antidepressant drugs or hormonal medicines such as contraceptive pills. These molecules can be small and difficult for treatment plants to remove without expensive upgrades.

Inorganic water pollutants include:

- Acidity caused by industrial discharges (especially sulfur dioxide from power plants).
- Ammonia from food processing waste.
- Chemical waste as industrial by-products.
- Fertilizers containing nutrients--nitrates and phosphates—which are found in storm water runoff from agriculture, as well as commercial and residential use.
- Heavy metals from motor vehicles (via urban storm water runoff) and acid mine drainage.
- Secretion of creosote preservative into the aquatic ecosystem.
- Silt (sediment) in runoff from construction sites, logging, slash and burn practices or land clearing sites.

Macroscopic pollution – large visible items polluting the water – may be termed "floatables" in an urban storm water context, or marine debris when found on the open seas, and can include such items as:

- Trash or garbage (e.g. paper, plastic, or food waste) discarded by people on the ground, along with accidental or intentional dumping of rubbish, that are washed by rainfall into storm drains and eventually discharged into surface waters.
- Nurdles, small ubiquitous waterborne plastic pellets.
- Shipwrecks, large derelict ships.

The Brayton Point Power Station in Massachusetts discharges heated water to Mount Hope Bay.

Change in Temperature

Thermal pollution is the rise or fall in the temperature of a natural body of water caused by human influence. Thermal pollution, unlike chemical pollution, results in a change in the physical properties of water. A common cause of thermal pollution is the use of water as a coolant by power plants and industrial manufacturers. Elevated water temperatures decrease oxygen levels, which can kill fish and alter food chain composition, reduce species biodiversity, and foster invasion by new thermophilic species. Urban runoff may also elevate temperature in surface waters.

Thermal pollution can also be caused by the release of very cold water from the base of reservoirs into warmer rivers.

Measurement

Environmental scientists preparing water autosamplers.

Water pollution may be analyzed through several broad categories of methods: physical, chemical and biological. Most involve collection of samples, followed by specialized analytical tests. Some methods may be conducted *in situ*, without sampling, such as temperature. Government agencies and research organizations have published standardized, validated analytical test methods to facilitate the comparability of results from disparate testing events.

Sampling

Sampling of water for physical or chemical testing can be done by several methods, depending on the accuracy needed and the characteristics of the contaminant. Many contamination events are sharply restricted in time, most commonly in association with rain events. For this reason "grab" samples are often inadequate for fully quantifying contaminant levels. Scientists gathering this type of data often employ auto-sampler devices that pump increments of water at either time or discharge intervals.

Sampling for biological testing involves collection of plants and animals from the surface water body. Depending on the type of assessment, the organisms may be identified for biosurveys (population counts) and returned to the water body, or they may be dissected for bioassays to determine toxicity.

Physical Testing

Common physical tests of water include temperature, solids concentrations (e.g., total suspended solids (TSS)) and turbidity.

Chemical Testing

Water samples may be examined using the principles of analytical chemistry. Many published test methods are available for both organic and inorganic compounds. Frequently used methods include pH, biochemical oxygen demand (BOD), chemical oxygen demand (COD), nutrients (nitrate and phosphorus compounds), metals (including copper, zinc, cadmium, lead and mercury), oil and grease, total petroleum hydrocarbons (TPH), and pesticides.

Biological Testing

Biological testing involves the use of plant, animal or microbial indicators to monitor the health of an aquatic ecosystem. They are any biological species or group of species whose function, population, or status can reveal what degree of ecosystem or environmental integrity is present. One example of a group of bio-indicators are the copepods and other small water crustaceans that are present in many water bodies. Such organisms can be monitored for changes (biochemical, physiological, or behavioral) that may indicate a problem within their ecosystem.

Control of Pollution

Municipal Wastewater Treatment

In urban areas of developed countries, municipal wastewater (or sewage) is typically treated by centralized sewage treatment plants. Well-designed and operated systems (i.e., with secondary treatment steps or more advanced treatment) can remove 90 percent or more of the pollutant load in sewage. Some plants have additional systems to remove nutrients and pathogens, but these more advanced treatment steps get progressively more expensive.

Nature-based solutions are also being used instead of (or in combination with) centralized treatment plants.

Deer Island Wastewater Treatment Plant serving Boston, Massachusetts and vicinity.

Cities with sanitary sewer overflows or combined sewer overflows employ one or more engineering approaches to reduce discharges of untreated sewage, including:

- Utilizing a green infrastructure approach to improve storm water management capacity throughout the system, and reduce the hydraulic overloading of the treatment plant.
- Repair and replacement of leaking and malfunctioning equipment.
- increasing overall hydraulic capacity of the sewage collection system (often a very expensive option).

On-site Sanitation and Safely Managed Sanitation

Households or businesses not served by a municipal treatment plant may have an individual septic tank, which pre-treats the wastewater on site and infiltrates it into the soil. This can lead to groundwater pollution if not properly done.

Globally, about 4.5 billion people currently do not have safely managed sanitation, according to an estimate by the Joint Monitoring Programme for Water Supply and Sanitation. Lack of access to sanitation often leads to water pollution, e.g. via the practice of open defecation: during rain events or floods, the human feces are moved from the ground where they were deposited into surface waters. Simple pit latrines may also get flooded during rain events. The use of safely managed sanitation services would prevent this type of water pollution.

Industrial Wastewater Treatment

Some industrial facilities generate wastewater that is similar to domestic sewage and can be treated by sewage treatment plants. Industries that generate wastewater with high concentrations of organic matter (e.g. oil and grease), toxic pollutants (e.g. heavy metals, volatile organic compounds) or nutrients such as ammonia, need specialized treatment systems. Some industries install a pre-treatment system to remove some pollutants (e.g., toxic compounds), and then discharge the partially treated wastewater to the municipal sewer system. Industries generating large volumes of wastewater typically operate their own treatment systems. Some industries have been successful at redesigning their manufacturing processes to reduce or eliminate pollutants, through a process called pollution prevention.

Dissolved air flotation system for treating industrial wastewater.

References

- Waterquality., ocean; floridakeys.noaa.gov, Retrieved 26 July, 2020
- Water-quality-monitoring-115, solutions: adcon.com, Retrieved 21 May, 2020
- Water-parameters, surface- water-quality, water-monitoring, water-monitoring-and-assessment, water; tas.gov.au, Retrieved 13 May, 2020
- Investigation of pollution incidents". Queensland government - department of environment and heritage proetection. 21 july 2016. Retrieved 1 august 2016
- Why-is-water-quality-monitoring-important, breaking-news, water-wastewater, news,: envirotech-online.com, Retrieved 8 January, 2020
- Hanaor, d.; michelazzi, m.; leonelli, c.; sorrell, c.c. (2011). "the effects of firing conditions on the properties of electrophoretically deposited titanium dioxide films on graphite substrates". Journal of the european ceramic society. 31 (15): 2877–2885. Arxiv:1303.2757. Doi:10.1016/j.jeurceramsoc.2011.07.007

Permissions

All chapters in this book are published with permission under the Creative Commons Attribution Share Alike License or equivalent. Every chapter published in this book has been scrutinized by our experts. Their significance has been extensively debated. The topics covered herein carry significant information for a comprehensive understanding. They may even be implemented as practical applications or may be referred to as a beginning point for further studies.

We would like to thank the editorial team for lending their expertise to make the book truly unique. They have played a crucial role in the development of this book. Without their invaluable contributions this book wouldn't have been possible. They have made vital efforts to compile up to date information on the varied aspects of this subject to make this book a valuable addition to the collection of many professionals and students.

This book was conceptualized with the vision of imparting up-to-date and integrated information in this field. To ensure the same, a matchless editorial board was set up. Every individual on the board went through rigorous rounds of assessment to prove their worth. After which they invested a large part of their time researching and compiling the most relevant data for our readers.

The editorial board has been involved in producing this book since its inception. They have spent rigorous hours researching and exploring the diverse topics which have resulted in the successful publishing of this book. They have passed on their knowledge of decades through this book. To expedite this challenging task, the publisher supported the team at every step. A small team of assistant editors was also appointed to further simplify the editing procedure and attain best results for the readers.

Apart from the editorial board, the designing team has also invested a significant amount of their time in understanding the subject and creating the most relevant covers. They scrutinized every image to scout for the most suitable representation of the subject and create an appropriate cover for the book.

The publishing team has been an ardent support to the editorial, designing and production team. Their endless efforts to recruit the best for this project, has resulted in the accomplishment of this book. They are a veteran in the field of academics and their pool of knowledge is as vast as their experience in printing. Their expertise and guidance has proved useful at every step. Their uncompromising quality standards have made this book an exceptional effort. Their encouragement from time to time has been an inspiration for everyone.

The publisher and the editorial board hope that this book will prove to be a valuable piece of knowledge for students, practitioners and scholars across the globe.

Index

www.ingramcontent.com/pod-product-compliance
Lightning Source LLC
LaVergne TN
LVHW080619210926
848485LV00003B/184